OXFORD MEDICAL PUBLICATIONS

Advances in international maternal and child health
Volume 2

Advances in international maternal and child health

Volume 2

Edited by

D. B. JELLIFFE and E. F. P. JELLIFFE

Division of Population, Family and International Health,
School of Public Health, University of California, Los Angeles

OXFORD
OXFORD UNIVERSITY PRESS
NEW YORK TORONTO
1982

Oxford University Press, Walton Street, Oxford OX2 6DP

London Glasgow New York Toronto
Delhi Bombay Calcutta Madras Karachi
Kuala Lumpur Singapore Hong Kong Tokyo
Nairobi Dar es Salaam Cape Town
Melbourne Auckland
and associate companies in
Beirut Berlin Ibadan Mexico City Nicosia

British Library Cataloguing in Publication Data

Advances in international maternal and
   child health.—(Oxford medical publications).
   —Vol. 2
   1. Pediatrics—Periodicals
   618.92'0005   RJ1

ISBN 0-19-261370-7

Set by The Thetford Press, Norfolk.
Printed in Great Britain by
Richard Clay Ltd, Chaucer Press, Bungay

Preface

It has often been remarked that in considerations of maternal and child health (MCH), the emphasis is more often on the child rather than on the mother. Indeed, recent surveys of the literature have indicated this disproportionate attention. For example, during the past 10 years appropriate technology for the health and nutritional surveillance of young children has received much more investigation than have methods for mothers. However, the section by Dr Kusum Shah represents pioneer work in Kasa, Bombay State, in devising appropriate technology for use in perinatal supervision by primary health care workers in rural areas in less developed countries (p. 1).

This volume also highlights the dyadic interaction between mother and child in maternal nutrition, especially in pregnancy, and in immunity against neonatal infection, as discussed by Dr Charlotte Neumann (p. 16).

The significance of breast feeding in this dyadic interaction cannot be overestimated, as clearly shown by the continuing flood of information publications, and research. Recent emphasis is given in the account of the 'precipitous decline' in breast feeding in the Arab Gulf States contributed by Dr Jamal Harfouche (p. 203), which can be regarded only as a public health reverse rather than an advance—as, indeed, indicated by the rising risk of diarrhoeal diseases and malnutrition in infants.

With pressures to bottle feed developing in varying degrees in peri-urban areas in many parts of the Third World, programmes to halt or mitigate this decline need to be much to the forefront of plans for rational infant nutrition. Emphasis is increasingly given to improving the education of health professionals and making services for pregnant women more conducive for breast feeding. Legislation to facilitate breast feeding for working women is also under much debate. In addition, the need for control of the aggressive promotion of formulas in developing countries is well recognized. By the time this volume was published, the much-debated WHO *Code of Marketing of Breast Milk Substitutes* had been approved by the World Health Assembly in Geneva, May, 1981.

The section by Dr Pierre Borgoltz (p. 158) is therefore important as it gives insight to usually unsensitized health professionals into the promotional realities of the infant formula industry, rather than their public relations images. The need for this awareness become greater as marketing increasingly focuses on health professionals.

The role of 'public health', as an aspect of 'medicine' has in the past often tended to be considered in a cultural vacuum. Fortunately, this has changed somewhat recently, when it was realized increasingly that any form of medical or health activity represents a form of cultural interaction, and that all patterns of traditional medical, nutritional and other beliefs contain both scientifically sound and unsound components. Because of this recognition, serious attempts are nowadays being given to try to see how traditional indigenous medical systems and western technological medicine systems can be brought together to reinforce one another, rather than to act in opposition and competiton.

Dr Robert Bannerman, who has been in charge of the WHO Traditional Medicine Program for some years, gives his views on integrating traditional and modern health systems (p. 28). Likewise, the editors summarize a practical approach to this type of cultural intergration and synthesis, with special relation to primary child health care (p. 50). In particular, in this interaction emphasis needs to be given to the harmful effects of Western medical cultural practices and concepts, as well as, in the past, focusing only on the indigenous culture.

With rising inflation, decreasingly adequate food supplies, massive moves to urban areas and population dislocations resulting from man-made or natural disasters, the nutritional situation for young children in many parts of the world had deteriorated, although, happily, improvements have taken place in limited numbers of developing countries.

A major issue remains the organization of low-cost, nutritional surveillance to try to develop an early-warning system and to direct resources to areas of priority need before the onset of serious ill consequences. Some of the information required for nutritional surveillance can be obtained from continual collection, collation and analysis of data from existing agencies and institutions, such as the Meterology Service, the Ministries of Agriculture and Labour, etc. However, there will always be the need for complementary direct information on the nutritional status of physiologically vulnerable groups. With this in mind, the idea of 'focused national nutritional surveys' (FNNS) has been developed. Some of the results, problems, and shortcomings are outlined by Dr Alfred Zerfas and colleagues (p. 56). These include comments on such universal problems as sampling, cost, training and anthropometric reference levels, and the need for epidemiological and statistical guidance modified by practical feasibility.

Protein-energy malnutrition remains the world's largest nutritional problem, but as Dr Mehari Gebre-Medhin notes (p. 87), rickets remains an unsolved public health problem in many parts of the world, both in less-developed and urban industrialized countries. Indeed, in some areas, there is evidence that rickets may be becoming an increasingly common condition. Also, partly in the field of nutrition remains the curious enigma of bladder stones. These often used to be seen in many parts of the world, including the temperate zone. However, nowadays they seem to have high incidence rates only in some tropical regions. Parts of Thailand seem to especially affected, as noted by Dr Aree Valyasevi (p. 98), who outlines the present situation.

Current views on two widely differing issues of considerable social significance have been described by Drs Yngve Hofvander and Claes Sunderlin, and by Dr Alan Norton. The first concerns the world-wide questions raised by 'transnational adoptions' (p. 111), with their inherent cultural and psychological overtones. In addition, Dr Alan Norton describes the curious and apparently bizarre phenomenon of epidemic hysteria in schoolchildren, which has been described in many parts of the world (p. 117), and which plainly has historical similarities to other psychogenic epidemics—for example, the 'dancing sickness' of the Middle Ages in Europe.

Lastly, but importantly, the question of education is emphasized, both in the sense of supplying information and/or inducing behaviour change. This plainly should be a major public health activity in all parts of the world. Some of the newer views on nutrition education are brought together (p. 133), while Drs David Morley and Felicity Savage describe the truly significant and remarkable activities of TALC (Teaching Aids at Low Cost) during recent years (p. 148).

The content of this volume of *Advances in international maternal and child health* indicates clearly that despite the differences between so-called developing and industrialized countries, and despite special problems and particular considerations in different areas of the world, many universal themes run through such considerations. This realization can be helpful when defining and implementing programmes and policies to improve the health and nutrition of mothers and young children anywhere.

Los Angeles   DBJ
March 1982   EFPJ

Contents

List of contributors

R. H. O. Bannerman
Formerly WHO, Geneva, Switzerland

Pierre A. Borgoltz
UN Centre for Transnational Corporations, United Nations, New York, USA

B. Browdy
Division of Biostatics, School of Public Health, University of California, Los Angeles, USA

W. D. Clay
Peace Corps, Nairobi, Kenya

Mehari Gebre-Medhin
Department of Pediatrics, University Hospital, Uppsala, Sweden

Jamal K. Harfoude
Faculty of Health Sciences, American University, Beirut, Lebanon

Yngve Hofwander
Department of Pediatrics, University Hospital, Uppsala, Sweden

D. B. Jelliffe
Division of Population, Family and International Health, School of Public Health, University of California, Los Angeles, USA

E. F. Patrice Jelliffe
Division of Population, Family and International Health,
School of Public Health, University of California, USA

David Morley
Tropical Child Health Unit, Institute of Child Health, University of London

Charlotte G. Neumann
University of California, Los Angeles,
Schools of Public Health and Medicine, Los Angeles, USA

Alan Norton
London, UK

Felicity Savage
Tropical Child Health Unit, Institute of Child Health, University of London

Kusum P. Shah
Formerly Department of Obstetrics, Grant Medical College, Bombay, India

I. J. Shorr
Division of Nutritional Sciences, Cornell University, Ithaca, New York

Claes Sundelin
Department of Pediatrics, University Hospital, Uppsala, Sweden

Aree Valyaseri
Institute of Nutrition and Faculty of Medicine, Ranvakhikodi Hospital, Mahidol University, Bangkok, Thailand

A. J. Zerfas
Division of Population, Family and International Health, School of Public Health, University of California, Los Angeles, USA

1 Appropriate technology and perinatal care: the Kasa experience

KUSUM P. SHAH

It is a sad fact that modern scientific technology has made a minimal contribution to reducing the maternal and perinatal mortality and morbidity in the remote areas of developing countries, where 60–80 per cent of the population of these countries reside. Seven out of ten maternal deaths in rural areas are due to preventable causes like haemorrhage, severe anaemia, sepsis, obstructed labour, and toxaemia.[1] Similarly, the vast number of perinatal deaths associated with obstructed labour, infection, and maternal mal-nutrition could be prevented if timely health care were provided.[2,3] The health-centre-oriented health services manned by doctors and nurses have limitations in reaching all pregnant women and newborns at grass-roots level due to various reasons. Taking as an example India: here a primary health centre may be supposed to serve a population of about 95 000 in 110–120 villages; on an average, there are 2–3 doctors. One-fifth of the centres have only one doctor each. Eighty per cent of the patients attending the outpatient clinic come from an area within a radius of 4.8 km, though there are some villages at a distance of 40 km from a health centre. A primary health centre has seven to eight sub-centres, each staffed by one Auxiliary Nurse Midwife (ANM) who has to provide services for a population of about 12 000 in approximately 15 villages. In an area served by one ANM, 420 women deliver and about four of these die as a result of childbirth every year. During the same period, there will be 15–40 perinatal deaths. In practice, an ANM serves only a limited section of the population in her area.[4]

The experiments on alternative approaches in health service delivery affecting primary health workers have shown wider coverage of the popula-tion and very high utilization rates of health services resulting in a remarkable reduction in maternal, perinatal, and infant mortality and morbidity.[5] After the WHO Alma Ata declaration in 1978,[6] many countries of the world have accepted the approach of primary health care and decided to put their efforts and resources to provide health-for-all by the year 2000.

The care of women during pregnancy is of utmost importance in influenc-ing the perinatal outcomes. In the rural community situated some kilometres from Bombay city served by the Health Unit, Palghar, District Thane, India, the maternal mortality rate was 7.8 per thousand in unbooked and 0.45 per thousand in booked cases. There was no death among the women who were attending antenatal clinics regularly. In a teaching hospital in Bombay, the maternal mortality rate was ten times higher in non-registered patients than in those registered. Similarly, the perinatal, early neonatal, and neonatal death rates for babies of booked mothers were about three times lower than those of unbooked mothers. The difference was more than four times in those who booked and regularly availed themselves of antenatal care. The incidence of birth weight of 2.5 kg or below at term was 22.1 per cent in booked mothers as compared with 62.1 per cent in unbooked mothers (Table 1.1).[2,3]

TABLE 1.1   *The effect of antenatal care on maternal and perinatal mortality rates and birth weights at the Rural Health Unit, Palghar, District Thane*

Indicator	Status of contracts	
	Registered	Non-registered
Maternal deaths (per thousand)	1	19
Maternal mortality rate (per thousand)	0.45	7.8
Perinatal deaths (per thousand)	193	186
Perinatal mortality rate (per thousand)	31.1	90.0
Birth-weight:		
2.5 kg or less (%)	22.1	62.1
Above 2.5 kg (%)	77.9	37.9

Until recently it was not possible to provide antenatal care for all the pregnant women owing to the limited trained medical manpower and resources, the long distances, and the problems of transport. Some of these problems will now be solved as primary health workers would be in a position to approach these women. Unless these health workers and their supervisors are properly trained in the skill of community diagnosis and management, the benefits of their presence will be negligible. The community management at the level of primary contact is entirely based on the community diagnosis by the health workers, who have a limited educational background; hence these workers should be equipped with appropriate technology for diagnosis and management. The utilization of technology would be maximized if the tools are simple, handy, easily portable in the shoulder bag of the worker, and, as far as possible, indigenous rather than imported. These instruments should be durable and cheap.

The development of appropriate technology should be based on the common problems of the community in the Third World. Many of the health problems of pregnancy and perinatal period are common, hence there are considerable chances of using the technology of one country in another. The usefulness of the tools devised for the field should be scientifically proved. After the pilot trial, the health workers, the staff from the health centre, and other health personnel should be trained in the correct use of the tools, the interpretation of the findings, and the line of action to be taken. Students of nursing and medicine should also be oriented in these devices so that they can appreciate the importance and usefulness of the technology.

The Kasa approach

The following appropriate technology for care of mothers and children was devised, tested, and evaluated at the Kasa project in Thane district, India.[7-10]

A. Appropriate technology for community diagnosis
   (i) mother's card
   (ii) technique to screen for stunted women
   (iv) anaemiometer
   (v) sling for weighing newborns
B. Appropriate technology for community management
   (i) 'Tiny' delivery kit
C. Appropriate technology for communication
   (i) Pigeonogram

A. Appropriate technology for community diagnosis

(i) *Mother's card*[7]

A comprehensive card has been designed for a married woman in the reproductive period to provide information on her menstrual status, pregnancy period if she is pregnant, state of nutrition before and during pregnancy, whether at risk during the pregnancy, immunizational status against tetanus, due month of confinement, labour, early post-partum period, whether breast feeding her infant, and whether using family planning measures. It is also a health education tool. The card could be used for the entire reproductive period of a woman, and there is a provision for the recording of four pregnancies during that time. There is a thick card—printed in the local language —to be retained by the woman in a plastic bag; and a duplicate card—on thin, white paper—which could be kept with the primary health worker or at the Health Centre, where it could be referred on follow-ups or for monitoring.

On the first panel of the threefold card, the health worker is supposed to enter the name of the village, date of completion of the card, name of the woman and her address, her present age, the age at menarche, and age at marriage. Sometimes the worker has to obtain the help of others in order to ascertain the approximate age of the woman by indirect evidence such as the age of her eldest child. The woman's weight, taken by the health worker on a portable spring balance scale, and height, measured against a straight wall or the trunk of a tree, are recorded. The information of her previous obstetric performance is entered in a tabular form in the first panel (Fig. 1.1). Then the sixth, or back-central, panel of the folded card is completed. The table on this panel has columns for the months of the year, and age in years from 14 to 44.

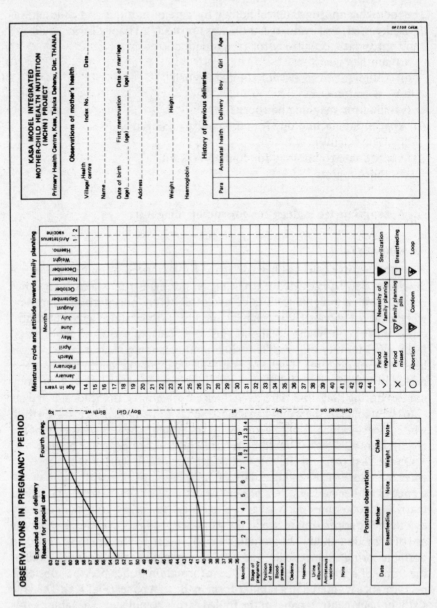

Fig. 1.1. Mother's card, panels 1, 5, and 6.

At the bottom of the table, various symbols are printed which indicate whether the woman is menstruating, has missed her period, is lactating, has had an abortion, is taking oral contraceptives or using condoms, or has an intrauterine device (IUD), and whether breast feeding. The health worker has to fill in the specific information in the column against her age and the calendar month (Fig. 1.1).

If the woman misses her period, the history is confirmed the next month, and the second or subsequent panel of the folded card is used during that pregnancy period (Fig. 1.2). The health worker records weight and some observations such as swelling of legs, whether at risk, and immunizational status against tetanus. The card is completed every month during the first 7 months of pregnancy, and fortnightly during the eighth and ninth months. The weight is plotted on the graph against the month of pregnancy. Two weight curves on a pregnancy page depict the weights during pregnancy. Those women whose weight line is near the upper curve are likely to deliver babies of 3.1 kg and above. Those below the lower line are at risk of giving birth to low-birth-weight babies.[11-13] Information was available at the Kasa Project about women who delivered babies having a birth weight of 2.8 kg or above and on their pre-pregnancy weight and weight during various months of pregnancy. The mean pre-pregnancy weight of these women who delivered babies weighing 2.8 kg or above at birth was 40.2 kg and the mean gain in weight during pregnancy was 5.0 kg. They gained 0.6 kg by the fifth month.

These lines and the woman's weight curve are included on the card to educate her and her family members, and in an attempt to help her to attain a weight between the two lines so that her baby will have a good birth weight. The urine of pregnant women with swollen legs is examined for albumin, either by the health worker or her supervisor, the nurse. The nurse also palpates the abdomen, estimates the haemoglobin of some women, and ensures that the two doses, or a reinforcement dose, of tetanus toxoid is given during pregnancy and entered on the card as well as on the duplicate. The blood pressure of those women with swollen legs and albumin in their urine is checked by a medical officer. At the top of these pages of pregnancy records the worker enters the parity of the mother, and the nurse puts down the expected date of confinement. Against the at risk column, any of the following indicators for at risk,[14] if present, are mentioned:

—weighs 38.0 kg or less before pregnancy, or weighs 42.0 kg or less at the 24th week of pregnancy;
—is less than 145 cm in height;
—is severely pale;
—has a child from a previous delivery who weighed less than 2 kg or was very small;
—has swollen legs;

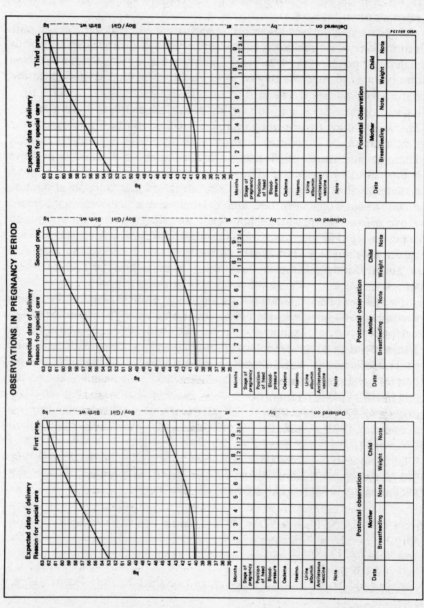

FIG. 1.2. Mother's card, panels 2, 3, and 4.

—is primiparous;
—is below 18 or above 35 years old;
—has a history of abortion or still-birth during a previous pregnancy;
—during a previous pregnancy lost her child within one month;
—is carrying her fifth or later child.

All at-risk pregnancies should be shown to the nurse or medical officer. Entries of the date of delivery, outcome of pregnancy, sex of child, and birth weight are entered on the same page. Provision for the recording of three or more pregnancies is made on the third, fourth (Fig. 1.2), and fifth panels (Fig. 1.1) of the folded card.

Experience gained from using the card. The cards have been used since 1973 in India, and were introduced in Somalia early in 1981. At the Kasa Model Integrated Mother–Child Health-Nutrition Project in India, 27 part-time social workers with a limited education of 4–10 years of schooling were able to complete 5263 cards of 88.2 per cent of the eligible women in 60 programme villages. It is surprising that even in the most remote villages there were records on the number of women menstruating, those who have missed their cycles recently, the number of women in various months of pregnancy, and those who were due to deliver during the next month. The information was also available on those who were at risk, those who had had abortions and still-births during the month, and those who had received tetanus vaccination. The records provided details on pre-pregnancy weight and weight gain during pregnancy, names of those who delivered by trained or untrained traditional birth attendants or ANMs, and if any had complications of delivery (Table 1.2). The menstrual registry prepared from these data provided information on the number of couples eligible for sterilization, termination of pregnancies, menstrual regulation, or any other contraceptive appliances. The satisfactory coverage was due to the widespread domiciliary coverage and the ability of the young male workers to gather information on the menstrual history with the help of either traditional birth attendants or their own female family members.

The cards reinforced the family planning programmes in the area. During 1974–5, only 74 sterilization operations were performed in Kasa. In 1975 and 1976, about 750 women and 1860 men were motivated to undertake the operation. The workers identified, on an average, two pregnancies out of 7–8 as at risk at any given time in a village with a population of about 900, and they were able to intervene in cases of severe malnutrition and severe anaemia by giving nutrition, education, and supplements and iron–folic acid tablets. These and other at-risk cases were referred or reported to the nurses or medical officer. The mothers kept the records safely so that they could be produced whenever needed. It was a very common sight to see children and women coming out from their houses with polythene bags containing their

TABLE 1.2 *Reporting design on menstrual registry for community information from 58 villages of the Kasa Project, District Thane; November, 1976*

Total No. of women's cards	Pregnancies								Outcome of pregnancies						Information on family planning			
										Labours conducted by:					Sterilized		Using	
	First tri-mester	Second tri-mester	7th month	8th month	9th month	Abor-tions	SB	Total preg-nancies	At-risk preg-nancies	Trained attend-ants	Un-trained attend-ants	Total births	Maternal deaths	Perinatal deaths	Male	Female	Loops	Con-doms
5263	54	138	60	38	57	10	Nil	357	36	Nil*	28	28	Nil	NR	622	11	62	49

NR = Not recorded
SB = Still birth
* Traditional birth attendants were trained later

yellow and pink cards during a visit from the health worker. The cards were found to be simple, very useful, informative, action-oriented, and to serve many purposes. They are comprehensive, but simple for the village-level worker. The card, its duplicate, and the polythene bag, costs about US $0.15 and last for about 10–15 years. The bag could also be used for the growth cards of young children in the family.

All the obstetric departments of teaching institutions, many maternity hospitals, and health centres and sub-centres have records for antenatal and postnatal care. However, at times these records are often incomplete, even though they are filled out by trained medical personnel. They serve as complex clerical records and are cumbersome for providing clear-cut information for a timely line of action. They do not provide any education to the women to whom they refer; whereas with the help of the mother's cards and similar growth cards for children it is possible to cover the entire vulnerable community, i.e. children under 5 years of age and married women, and provide timely curative, preventive, and promotional services by the health worker. It also simplifies the assignments of auxiliary nurses who have to examine and manage the at-risk pregnancies or refer high-risk cases to the doctors. It minimizes waiting time before examination. The card becomes a useful guide to the medical officer for the referred cases and it is an excellent source of detailed information for research in the community.

(ii) *Tricoloured arm tape for assessing maternal nutrition*

It is fairly well established that the incidence of low birth weight is higher in very poorly nourished women. However, it is often not feasible to carry out accurate body weight measurements in rural areas, and a simple field technique to assess the nutritional status of a woman does not exist. A significant correlation has been observed between arm circumference and the weight of women of child-bearing age, irrespective of precise age, parity, and height. This suggests that arm circumference can be used as an indicator of body weight[8] and may have potential for use during pregnancy to identify at-risk pregnancies, culminating in the delivery of a low-birth-weight baby. The average weight gain during pregnancy in developing countries does not leave much margin for increase in adipose tissues after allowing for the weight of a full-term fetus, expansion of blood volume, placenta and liquor amnion. It would thus be reasonable to expect that the arm circumference may not be influenced much by pregnancy, especially in poorly nourished women.

The observations on body weight and arm circumference in a woman have been incorporated on a plastic tape. It was observed that those women who had an arm circumference of 20.8 cm or less weighed less than 37.5 kg and were, therefore, considered severely malnourished. Those who had an arm circumference between 20.9 and 22.8 cm and weighed between 37.5 and 45.0 kg were grouped as moderately malnourished, and those with arm circumferences above 22.8 cm, weighing 45.0 kg or more, were grouped as mildly malnourished or normal. The critical cut-off points in grading of nutritional status were based on the previous studies demonstrating the interrelationship between maternal weight and low birth weight in the newborn.[4,12,13] Tricoloured arm tapes with bands of pale yellow at 18.5–20.8 cm from the end, orange from 20.8 to 22.7 cm, and bright red for over 22.7 cm were produced. The colours indicate severe, moderate, and mild malnutrition or normal nutrition, respectively (Fig. 1.3). The villagers in India relate red with blood, i.e. good health, and they are not familiar with red traffic signals indicating danger. Hence, a red-coloured band represents good nutrition. This simple tricoloured tape can be used by primary health workers to detect malnourished, at-risk pregnant women for nutrition supplementation whose arm circumference falls within the pale yellow band.

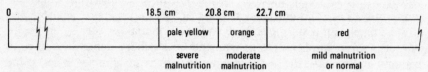

FIG. 1.3. Tri-coloured arm tape.

(iii) *Technique to screen for stunted women*

Stunted women may be at risk for delivering low-birth-weight babies. The height measurement of a pregnant woman was taken with a view to screening for stunted growth. The technology for identifying such a woman can be simplified by fixing a beam or plank of wood at the critical height from the ground or by finding a thick, straight, but horizontal branch of a tree at that height. The primary health worker asks pregnant women to pass under it without bending. Those who could pass are considered as stunted. At the time of taking the measurements, women should be bare-footed. This is a simple technique and has been found to be widely effective.

(iv) *Anaemiometer*

Severe anaemia in pregnancy is one of the major causes of maternal death and low birth weight. It is difficult to estimate haemoglobin concentrations without taking a blood sample, which only exasperates the problem. This technique is a study of haemoglobin values and the colour of conjunctiva; a simplified scale with three shades of red has been produced to help the primary health worker in diagnosing mild, moderate, and severe anaemia by comparing the different colourbands with the colour of the conjunctiva. The accuracy of these bands is in the range of ± 1.0 g of haemoglobin, and corresponds to 83.4, 78.0 and 71.9 per cent in severe, moderate, and mild anaemia respectively (Fig. 1.4).[4]

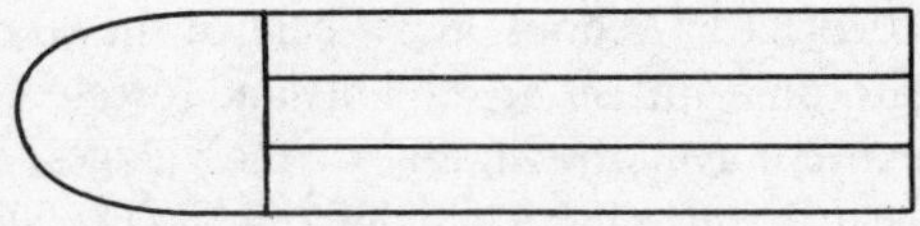

Colour bands representing:

mild anaemia

moderate anaemia

severe anaemia

FIG. 1.4. The Anaemiometer.

(v) *Sling for weighing newborns*

Birth weight has been suggested as one of the most important developmental indicators. However, a vast number of the babies in developing countries are delivered at home by traditional birth attendants and are not weighed. It is often a social custom not to take the newborns outside their homes for a month. The part-time social workers who weigh the under-fives with a spring balance can also weigh the newborns in their homes. The weighing is made possible using a specially devised sling prepared from a rectangular-shaped thick cloth of 1.2 by 0.7 m with cloth-hooks at the four corners. These can be made locally. The sling is also useful for weighing small infants who do not hold up their heads.

B. 'Tiny' delivery kit

At the beginning of this project, 98.8 per cent of deliveries were conducted by untrained traditional birth attendants (TBA) who were reluctant to undergo a 2-month training course.[4] They expressed their shyness to conduct deliveries with the UNICEF TBA kit, if it was provided for them. Moreover, the kit is big and cumbersome to carry. The TBAs are generally of a lower social stratum in Indian villages and they fear people's criticism when they walk into the village carrying a modern medical tool. Puerperal sepsis and neonatal and maternal tetanus were common causes of deaths in the district. Ninety-nine out of 134 TBAs in the Kasa Project villages underwent four days' training and were provided with 'tiny' delivery kits: a small plastic box contained sterilized blade, twine, gauze pieces, cotton swabs, and tincture of iodine. After a delivery, the TBA replenishes these items at the primary health centre. The kit is handy and easy for the TBA to carry at the waist, in a fold of the saree. The TBAs readily used the kits. The cost of the kit initially, was US $0.20, and refilling per delivery cost US $0.04.

C. Appropriate technology for communication

(i) *Pigeonogram*

A large number of villages in developing countries are unapproachable owing to difficult terrain and lack of transportation. In emergencies, the primary health worker may need the assistance of a nurse or doctor from the health centre which is generally at a walking distance of a few hours. In such situations, an appropriate technology of communication of messages was experimented with at the Kasa Project.[4,15] About 70 per cent of the 60 programme villages were not on a bus route and the area is hilly and forested. No telephone or telegraph facilities were available in any of the villages. The farthest village in the primary health centre service area from Kasa is 50 km by road, but only 30 km as the crow flies. A good homing pigeon could fly that distance in less than 30 minutes.

Rearing the pigeons. A loft measuring $2 \times 2 \times 2.5$ m was made for housing about 50 birds. Small-gauge wire netting was used so that rats could not enter. The loft had only one door, size 2×1.3 m, and one window, size 20×20 cm, by which the birds entered. Individual nesting boxes of 30×20 cm were hung on the wall, 1 m above the ground, and perches were provided for the birds to sit on when they were out of their nest boxes. Only two persons were assigned to look after the birds. To prevent the spread of diseases, no other kinds were kept in the loft.

Mixed grains were dispensed from wooden trays kept on raised surfaces. The water tray was kept away from the food tray, and water was changed 2–3 times daily. A sand/grit tray was kept in the loft, as pigeons eat sand or grit to help digestion. These trays were removed after dark. Chopped raddish leaves were fed once or twice a week in order to give the birds minerals and vitamins. A weak or small orphan bird was hand-fed 2–3 times a day with kneaded wheat flour until it could pick the grains by itself. A watch was kept for hawks, falcons, cats, or dogs if found circling around the loft.

Training the pigeons. Only healthy, fully developed pigeons accustomed to the area were taken for training. The birds were only flown during the day up to 3–4 hours before dark so that they returned to the loft by sunset. 4–6 birds at a time were taken out in a basket and released so that they flew in a group. The time of release and arrival was noted, along with the distance flown for checking the speed as well as any delay. Before being released, the birds were fed one-fourth of a day's requirements plus water if the distance to be flown was long. Food and water were kept in the loft so that the birds would fly directly there after their arrival from the training flight. White birds were not trained, but used for mating with plain, dark-coloured birds, as hawks can catch a white one by easily noticing when it is in flight. When a pair were breeding, they were not taken for training, and not distrubed for any reason.

Pigeons were trained in the mornings. For the first toss, the distance was not more than 200 m, and it was repeated from all four directions at about 3-hourly intervals. The birds were tossed in the air, pointing their heads in the direction of the loft. The second toss was made from 500 m, and repeated from all directions at 3-hourly intervals. For the third, fourth, and fifth tosses, distances were 1 km, 2 km, and 3 km, respectively. The intervals between repeat tosses in different directions were more than 6 hours. The sixth toss was from 8 km and the seventh from 15 km, and these were repeated after an interval of 24 and 48 hours, respectively. Eighth and ninth tosses from 20 km and 30 km were repeated in all directions at intervals of 72 hours. For longer flights, strong and healthy pigeons were used. The distance was increased by 20 km at intervals of 4 days until the full distance had been covered. Thereafter, the bird was flown every week.

When tossing a bird, a weight of 3 g was tied to its leg so that when a capsule was tied it would not shy off. The weight was removed as soon as the bird entered the loft. After the ninth toss, the bird was flown from 10 km with an empty plastic capsule tied to its leg with a button, so as to get accustomed to it.

The pigeon as a messenger. It took about 3–4 months to train young pigeons. There were seventeen pigeons to start with, and within 2 years these had multiplied to fifty-six. Once there was a catastrophe. A stray dog started attacking the birds and killed thirty-five of them. However, after changing the position of their loft, and keeping a stricter vigilance, the remaining pigeons survived, and by 20 months the number grew to sixty-seven.

Two birds, without any consideration of sex, were kept caged in a selected village with a difficult approach until release by the primary-health-care worker. In the project area there were fifteen villages with difficult approaches. The pigeons were transported to these villages in a Land-Rover, which was used primarily for distributing the monthly quota of nutritional supplements for the health workers' service areas. If the pigeons were not released until the Land-Rover's visit, then they were released on that day and replaced by a fresh pair. They were fed and watered in the cages 2–3 hours a day. After their release, they were rested for a month before they were sent out again to another village. The resting pigeons were given regular tossing flights to keep them fit.

In case of an emergency, one bird was flown by the health worker, and the other one was released only if medical aid had not arrived within 3 hours. The message was placed in a plastic capsule which was buttoned to the leg. On receiving the message, the medical officer and nurse rushed in the hospital vehicle with the medicine and a fresh pair of pigeons. The system was implemented for 3 years, and an average of 3–4 messages were received. The messages brought news about illnesses such as acute diarrhoea with moderate or severe dehydration, pneumonia, difficult labour, ante- or postpartum haemorrhage, and other medical emergencies. Messages were also received on the movement of some supervisors, indicating their location and that others should join them or provide some supplies, which were hence able to reach the proper site without wasting time in searching. The message also helped in correcting the position of nutritional supplements and drug supplies for a primary-health-care-worker's service area.

Registration of arrival of the messenger. Those residing in the house or neighbouring houses which were opposite the loft kept a watch for the arrival of any messengers. At the Narangwal Project, Punjab, India, at the site of the pigeon-post where the messenger-bird arrived, a stage was fitted with an electronic device, and as the birds landed a bell rang. This device is only possible when there is a continuous supply of electricity, particularly during day time.

The pigeons were used during day time all the year round, except during torrential rains. There are reports that pigeons can be used at night as their geomagnetic sense is very accurate. The cost in running this system of communication was negligible, as the pigeons were fed on mixed grains. However, as there was no budgetary provision for supplying food for birds or animals at the Primary Health Centres, the birds could not be fed for longer than 1 year after the Project came to an end, and their use became infrequent. The following were some of the investigations which it was proposed to study:

1 a system of intercommunication between the health-worker's village and the Primary Health Centre;
2 feasibility of sending ampoules, tablets and blood slides along with a message;
3 a better position for buttoning the capsule in order to carry maximum weight.

These studies could not be completed. However, the experiences mentioned here are promising and rewarding, particularly in a system of primary health care.

The appropriate technology for maternal and perinatal services described above has been extensively tried in the field for some years. To a great extent, the success of the Kasa Project was on account of appropriate and improvised technology suited to local conditions and to the part-time social workers. The techniques and tools developed and tried at Kasa Project do need further studies in different countries for adaptation and modification. The technology has many benefits and its introduction in primary health care is a *sine qua non* for reducing morbidity and mortality in mothers and young children.

References

1. Shah Kusum, P., De Souza, J. M., Sawardekar, D., and Aphale, R. V. 'Comparative study of maternal mortality in the rural community and city teaching institution'. *Ind. J. Public Hlth.* **15**, 81–83 (1971).
2. Shah, P. M., Udani, P. M., and Shah Kusum, P. 'Analysis of vital statistics from the rural community, Palghar: (i) fetal wastage and maternal mortality'. *Ind. Pediatr.* **6**, 595–607 (1969).
3. Shah, P. M., and Udani, P. M. 'Analysis of vital statistics from the rural community, Palghar: (ii) perinatal, neonatal and infant mortalities'. *Ind. Pediatr.* **6**, 651–688 (1969).

4. Shah Kusum, P. 'Organization of maternity services in India'. In *Maternity services in the developing world—What the community needs* (ed. R. H. Philpot) The Royal College of Obstetricians and Gynaecologists, London (1979).
5. Newell, K. W. (ed.) *Health by the people*. World Health Organization, Geneva (1975).
6. World Health Organization. *Primary health care*. Geneva (1978).
7. Shah Kusum, P. 'Surveillance card for married women for better obstetric performance'. *J. Obstet. Gynaecol. Ind.* **28**, 1015–20 (1978).
8. Shah Kusum, P. 'Appropriate technology in primary health care for better midwifery services'. *J. Obstet. Gynaecol Ind.* **30**, 109–14 (1980).
9. Tibrewala, S. N. and Shah Kusum, P. 'The use of arm circumference as an indicator of body weight in adult women'. *Baroda J. Nutr.* **5**, 43–6 (1978).
10. Shah, P. M. and Shah Kusum, P. 'Role of teachers from medical colleges in delivery of health care in rural areas. Development of appropriate technology, training and operational research'. In *Delivery of health care in rural areas* (ed. V. Kumar) India, PGI, Chandigarh (1978).
11. Shah Kusum P. and Shah, P. M. 'Relationship of weight during pregnancy and low birth weight. *Ind. Pediat.* **9**, 526–31 (1972).
12. Shah Kusum, P. and Shah, P. M. 'Factors leading to severe malnutrition. Relationship of maternal nutrition and marasmus in infants'. *Ind. Pediatr.* **12**, 64–7 (1975).
13. Shah Kusum, P. and Shah, P. M. 'Relationship of maternal nutrition and low birth weight'. *Ind. Pediat.* **16**, 961–6 (1979).
14. Shah, P. M. and Shah Kusum, P. 'Community diagnosis and management of malnutrition. A realistic approach to combat malnutrition at the grass-roots level'. *Food Nutr. (FAO, Rome)* **4**, 2–7 (1978).
15. Shah, P. M. and Shah Kusum, P. 'The Mother's Card: a simplified aid for primary health workers. *WHO Chronicle* **35**, 51–53 (1981).

2 Maternal nutrition and neonatal immunocompetence

CHARLOTTE G. NEUMANN

Introduction

Maternal malnutrition whether due to the early onset of frequent and closely spaced cycles of pregnancy and lactation and/or energy deficits due to poor basic diet, hard physical work, and the catabolic effects of repeated bouts of infection, can lead to various 'maternal depletion syndromes'. Not only are energy deficits incurred but often deficiencies of protein, iron, calcium, folic acid, and occasionally vitamin A and the B-complex vitamins.[1] Pregnancy outcome in such circumstances may be poor with low birth weight due to intrauterine growth retardation (IUGR). Prematurity is also of greater incidence in maternal malnutrition and often associated with IUGR. Increased neonatal morbidity and mortality with short-term complications and long-term sequalae may occur in a number of areas of function.[2]

The role of maternal malnutrition in fetal growth retardation is now well established. The role of maternal dietary supplementation during pregnancy in increasing birth weight (BW) has been documented mainly in disadvantaged and nutritionally deprived populations, but not in well-nourished groups of women.[3] Studies by Roeder and Chow in Taiwan,[4] Lechtig et al. in Guatamala,[5], Burke et al. in the USA,[6] and by Higgins et al. in Canada[7] have demonstrated the association of dietary supplementation during pregnancy and increase in birth weight. Post-mortem studies by Naeye et al.[8] document the association of deficient maternal intake and decreased body and organ size and weight in intrauterine-growth-retarded infants born to mothers of low socioeconomic status. Adverse effects of IUGR include poor physical growth,[9] subtle learning disabilities,[2] and problems with infection and immune function.[10-14] This chapter will deal mainly with the effects of maternal or gestational malnutrition presumably present in intrauterine-growth-retarded infants and the effects upon the developing immune system.

Immune function in intrauterine-growth-retarded (IUGR) experimental animals

A limited number of studies in humans document a link between maternal malnutrition and defective immune function in the neonate. However, well-controlled animal studies exist supporting this hypothesis.

A very brief review of animal studies furnishes useful clues to the human situation in terms of which nutritional deficiencies may be most detrimental.

These include energy and protein deficits, zinc, pyridoxine, folic acid, B_{12}, and choline deficiencies. Gephardt and Newberne[15] demonstrated in rats and mice the role of marginal 'lipotrope' (choline, methionine, folic acid, and B_{12}) deficiencies singly or in combination in fetal growth retardation with particular hypoplasia of lymphoid, thymic and splenic tissue and impairment of cell-mediated immunity. Subclinical 'lipotrope' deficiencies, but particularly folic acid and B_{12} deficiencies, are prevalent in pregnant women, particularly among the poor, vegetarians, and women who are alcoholic.

Pregnant rats rendered zinc deficient were found to have offspring with fetal growth retardation as well as decreased cell-mediated immunity.[16] Pyridoxine deficiency in pregnant animals has resulted in IUGR progeny with decreased thymic size, reduced lymphoctye function, and increased mortality from infection. Pyridoxine deficiency impairs nucleic acid synthesis and in turn protein synthesis, cell division, and repair.[17] Both of the above deficiencies are widespread in humans in many parts of the world.

Calorie and protein-deprived pregnant animals give birth to offspring that have thymolymphatic atrophy, reduced cell-mediated immunity and impaired antibody response to T-cell dependent antigens.[18] The depressed immune response is seen also in the first- and second-generation offspring.[18] This work has important implications for humans in face of the global nature of energy deficiency.

In summary, experimental animal work demonstrates that maternal deficiencies of calories, protein, iron, zinc, 'lipotropes' (particularly folic acid), and pyridoxine result in thymolymphatic hypoplasia and impairment of cell-mediated immunity in the offspring. Even first- and second-generation offspring may be affected.[18] The mechanisms of action whereby maternal malnutrition affects the immune system of the offspring, although not completely understood, points to reduction of all lymphoid tissue due to reduced DNA and protein synthesis. Hormones, particularly cortisol, at times of maternal stress may pass to the fetus via the placenta and adversely affect developing lymphoid tissue, as has been shown in animals and humans.[18,19] Extrapolation of animal data to humans must, however, be made cautiously, but has much relevance.

Immune function in intrauterine-growth-retarded (IUGR) infants

The maternal situation

For the most part, in studies of immunocompetence of IUGR infants, the mothers were not examined for nutritional status, but rather it was inferred that because the mothers were in poor socioeconomic circumstances, they were probably receiving poor diets, had suboptimal pregnancy weight gains, and an increased likelihood of infection. Even moderate protein-energy malnutrition (PEM) in pregnant women was found to be associated with significant reduction in placental size and fetal growth retardation.[20]

The role of toxaemia, hypertension, placental malaria, and the TORCH[1] infections are well-known contributors to the problem of low BW due to fetal growth retardation and have a higher incidence in low socioeconomic groups.[2] Viral infections and malaria can cause obliterative vasculitis.[2] In studies of Costa Rican infants, Mata demonstrated that increased IgM levels, a sign of intrauterine infection, were present in IUGR infants compared with controls of normal birth weight.[21] Any conditions reducing the vascularity and blood flow to the placenta and fetus with decreased placental size and a diminished supply of energy and nutrients result in a higher incidence of IUGR infants.

In studies of neonates, Chandra,[10] Ferguson *et al.*[11] and Neumann *et al.*[12] found that IUGR infants had depressed cell-mediated immunity, as ascertained by a number of parameters of cellular immune function. The mothers in one study were shorter and lighter, had a higher incidence of anaemia and concentrations of serum transferrin and albumin were slightly decreased compared with mothers of healthy normal full-weight infants. Studies of poor Indian women by Singh *et al.*[13] and Bhashkaram *et al.*[14] showed an increased incidence of IUGR among neonates with decreased cell-mediated immunity. It was presumed, but not documented, that problems of malnutrition and infection in the mothers accounted for the fetal growth retardation and decreased immunity in the offspring.

Reduced availability of nutrients secondary to poor maternal diet or a decrease in placental blood flow can retard the growth of fetal lymphoid organs, especially the thymus and the developing immune system, as in animals.[8] DNA synthesis is decreased and cell division and growth can be severely restricted.[18] Post-mortem examinations of IUGR infants born to families of low socioeconomic status with poor nutritional intake revealed involution of the thymus, decreased splenic size, and decreased organ size and cell number.[8]

In addition to maternal protein and/or calorie malnutrition, folic acid and zinc deficiency in humans[23] can result in IUGR with decreased lymphoid and thymic tissue.[22,23] The decreased availability of essential nutrients which cross the placenta can result in decreased cell number and function of lymphoid tissue. In humans, maternal short stature is highly correlated with IUGR offspring. A viscious cycle of nutritional deficiency *in utero*, low birth weight, impaired physical growth during childhood and short adult stature follows and the cycle is thus repeated in the second generation.[18]

Transient immunodeficiency of the normal neonate

Clincial observations suggest that human neonates have impaired resistance to infection.[24] The full-term normal newborn of normal birth weight exhibits

---

[1] TORCH: T = toxoplasmosis; O = other; R = rubella; C = cytomegalic inclusion disease; and H = herpes.

some immune defects in all spheres of immune function, although none are severe and all readily correct themselves with maturation.[24] The clinical relevance of each defect is not entirely clear. A brief review of these problems will serve as a background for the additional and more long-lasting problems of the intrauterine-growth-retarded infant.

Monocytes are decreased in number at birth in the normal-weight full-term newborn, with decreased random migration, chemotaxis, and inflammatory responses.[25,26] This may be responsible for the poor walling off of infections. Granuloctyes show marked chemotactic defects. The cells are stiff and do not deform normally. Phagocytosis is normal but microbial killing is prolonged, requiring a long incubation period, rather than being defective.[27,28] Normal killing eventually does occur in the presence of normal opsonic activity. However, under stress of severe infection, an opsonic defect becomes apparent. Although the levels of the classical and alternate pathway complement components and C_3 activators are 30–50 per cent of normal levels this is not critical unless IgG is greatly reduced, as in premature birth, and killing function is reduced;[28,29] nitroblue tetrazolium (NBT) reduction by granulocytes, a sign of intracellular killing, is decreased.[27]

As for humoral immunity, B cells are normal in number and percentage but functionally do not synthesize immunoglobulins except for IgM. There is poor B-cell conversion to plasma cells, presumably due to the suppressor action of neonatal T cells. Therefore, newborns are poor antibody producers. Preterm infants start off with low immunoglobulin levels, especially IgG, which improve with maturation.[30]

Cellular immunity is somewhat decreased. By 38 weeks' gestation, delayed cutaneous hypersensitivity (DCH) and E-cell rosette formation are present but decreased.[31] The percentage of T cells is slightly low but the total number is normal at birth. Upon stimulation with a mitogen and spontaneously, lymphocyte blastogenic transformation is high. Lymphocyte cytotoxicity and migratory inhibition factor (MIF), both measures of cell-mediated immunity, show reduced activity and lymphokine production (interferon) is slightly decreased.[32] In summary, protection against infection at birth is accomplished partly by transplacental transfer of maternal antibody (IgG and IgM) and by the anti-infective properties of colostrum and breast milk such as secretary IgA, lymphocytes, macrophages, lysozyme, etc.[33-35]

Problems of immune function and infection in IUGR infants

Humoral immunity. Humoral immunity is generally normal in terms of percentage and number of B cells in IUGR infants,[11-14] except in one study where B-cell number was reduced.[10] Levels of immunoglobulins are generally normal or increased, except where the IUGR infant is also pre-term.[10] Immunoglobulin IgG and IgE levels both correlate with gestational age.[36] IUGR infants born as a result of placental dysfunction may have low IgG levels because of decreased transfer, via the placenta.[36] In a study of rural Costa Rican infants, as mentioned above, increased IgM levels in cord blood of IUGR infants were presumably due to high infection prevalence while *in utero*.[21]

Antibody responses to tetanus and typhoid immunizations have been observed to be at protective levels in normal and IUGR infants. However, a greater portion of IUGR infants had lower post-immunization concentrations of tetanus and typhoid antibodies then did normal infants.[10] Seroconversion with oral polio vaccine (OPV) was seen in low-birth-weight infants, both pre-term and full term, but the mean antibody titre was lower at 4 and 8 weeks after immunization than in normal BW infants. Haemagglutination inhibition antibody (HAI) response to measles vaccine was found to be comparable in IUGR and normal infants[37] when cellular immunity is intact. However, Neumann *et al*.[12] found a higher percentage of non-responders (HAI titres 1:5 or 0) if cell-mediated immunity was impaired at birth.[12]

Decreased specific secretory IgA antibody levels are found in IUGR infants.[10] This reduced function of the secretory immune system allows for an increase in circulating antibody to food protein in IUGR infants.[10] This implies a failure of prevention of mucosal penetration in the IUGR infant, permitting absorption of dietary protein.

Complement levels. Although studies of IUGR offspring of malnourished dams (PEM) demonstrated low complement concentrations this has not been seen in humans. Studies of Indian and Kenyan infants with IUGR were found to have normal concentrations of complement components C_3 and C_4.[12,38]

Cell-mediated immunity (CMI). All studies of IUGR infants, whether term or preterm, show diminished CMI.[10-14] This is expressed by multiple para-meters such as decreased total lymphoctye count, reduced T-lymphoctyes as enumerated by E-rosette forming cells (RFC), and diminished delayed cutaneous hypersensitivity. IUGR infants given BCG at birth have a high per-centage of negative reactions to intradermal PPD (protein-precipitated derivative of tuberculin) after immunization.[12,39] *In vitro*-lymphocyte stimu-lation is normal or increased as there is increased spontaneous blastogenic activity among low- and normal-birth-weight neonates.[32] The range of depression of CMI is similar in all the studies, a moderate but significant degree of decreased function (Table 2.1). Rather than a threshold effect, a spectral or progressive depression of CMI is seen depending on the degree of IUGR as noted by birth-weight deficit. One study examined IUGR infants, not only below 2500 g but also between 2500 and 2800 g. The latter BW group had decreased CMI intermediate between controls and the smaller IUGR infant groups.[12] Other studies examining very-low birth-weight groups among the IUGR infants also found a diminishing degree of CMI function with decreasing birth weight.[14] The degree of IUGR is apparently the main determinant of CMI depression (Figs. 2.1 and 2.2).

TABLE 2.1 *Cell-mediated immunity in IUGR infants according to BW groups in several studies*

BW(g)	Ferguson et al.[11] % RFC*	Bhaskaram et al.[14] % RFC	Neumann et al.[12] % RFC	Singh et al.[13] % RFC
>2800	65.1±1.4	—	60.2±1.1	—
2501–2800	—	56.8±1.43‡	56.5±1.3	49.6±6.7†
2250–2500	—	52.3±3.26	52.5±7.2	—
1801–2250	49.2±2.0	36.3±3.01	49.8±5.2	54.4±7.8
1501–1800	—	—	44.0±8.3	23.9±4.2

* Mean±SE.
† BW groups 2600–3900 g.
‡ BW group >2500 g.

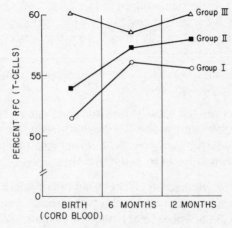

FIG. 2.1. Percentage RFC (T cells) in IUGR and normal infants at birth, 6, and 12 months. (Group I, BW < 2500 g; II, BW 2501–2799 g; III, BW > 2800 g.)

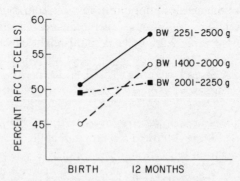

FIG. 2.2. Mean percentage RFC at birth and 12 months in IUGR infants < 2500 g BW, according to BW groups.

Normal catch-up growth helps, only in part, to restore the impaired immune function. Unlike post-neonatally acquired PEM, where CMI shows recovery early in the course of nutritional rehabilitation (often by 10–14 days)[40] the depressed CMI in IUGR infants is long-term, lasting from 1 year to at least 5 years, as found by Ferguson[41] and Chandra.[10]

Polymorphonuclear (PMN) leukocyte function. Intrauterine growth retardation is associated with a significant decrease in PMN leukocyte number in some studies. Intracellular killing is decreased and there is a decrease in NBT reduction.[10] Impaired mobilization and killing capacity of PMN leads to suboptimal tissue inflammatory response and host protection is diminished.[10] Also, macrophages may show similar diminished function.[26] Leukocyte metabolism, studied in the blood of mothers of IUGR infants and cord blood leukocytes, showed reduced oxygen uptake, reduced oxidase activity, a decrease in the hexomonophosphate shunt, and altered pyruvate kinase activity: all evidence of an energy deficit. It was concluded that leukocytes, however, did not necessarily show severe functional impairment relevant to host resistance to infection.[42]

Functional outcomes

Few studies have dealt with the functional outcomes or clinical importance of the immune problems in the IUGR infant. A group of IUGR infants and normal BW controls in Kenya were followed up to study immune function, response to immunizations, and morbidity.

Response to immunizations. The humoral response to immunizations in IUGR infants as mentioned earlier appeared to be adequate for tetanus, poliomyelitis, and typhoid but tended to produce lower post-immunization antibody concentrations than in infants of normal birth weight.[10,36] Those levels none the less were considered protective against clinical disease.

With regard to CMI status and immunization response, non-responders to pertussis and to measles immunization were found to have a significantly higher percentage of infants with reduced T cells than did vaccine responders[12,43] (Table 2.2). BCG immunizations given at birth resulted in

TABLE 2.2   *Vaccine response and cell-mediated immunity*

Vaccine	% Low rosettes
Measles:	
Non-responders (HAI*<1:5)	20.2
Responders (HAI≥1:5)	4.2
Pertussis:	
Non-responders (<1:10)	53.0
Responders (≥1:10)	37.1

* HAI = Haemaglutination inhibiton antibody.

significantly lower percentages of positive tuberculin reactions at 6 months after BCG in IUGR infants than in normal newborns, another expression of depressed CMI[12,39] (Fig. 2.3). Since depressed CMI is the main immunological problem in IUGR, this may decrease the effectiveness of certain T cell dependent immunizations in these infants, as against tuberculosis, measles, and pertussis.

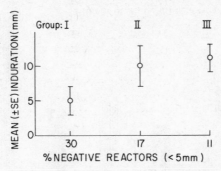

FIG. 2.3. Delayed cutaneous hypersensitivity in normal infants and in those with IUGR after BCG immunization at birth (5TU PPD at 6 months) (Group I, BW < 2500 g; II, BW 2501–2799 g; III, BW > 2800 g.)

Clinical infections and CMI. In examining morbidity in IUGR infants in whom immune function had been studied, crude clinical infection rates from 0 to 6 months of life were higher than in normal BW infants. Episodes of lower respiratory infection, oral moniliasis, and pertussis were higher in the IUGR infants with decreased percentages of T cells (50 per cent) compared with those with normal T cells (Fig. 2.4).[43]

In the same study, over the entire first year of life, double the percentage of IUGR infants had 10 or more clinical infections compared with the normal BW infants (26 *v.* 14 per cent). Forty-three per cent of the latter group had reduced T cells compared with 13 per cent of the normal infants. Between 7 and 12 months of age, otitis media, clinically diagnosed tuberculosis, and diarrhoea were more prevalent in the infants with decreased CMI than in infants with normal CMI. All of the infants with tuberculin (PPD) skin test reactions greater than 15 mm and positive chest X-rays had decreased T cells at birth.[43]

Concluding remarks

In conclusion, the health implications of depressed CMI and of the opsonic and leukocyte abnormalities in the IUGR infant include increased susceptibility and lower resistance to infection and diminished response to certain immunizations. The IUGR infant, already suffering from diminished CMI at birth, is at greater risk for postnatal malnutrition[9] and additional insult to its

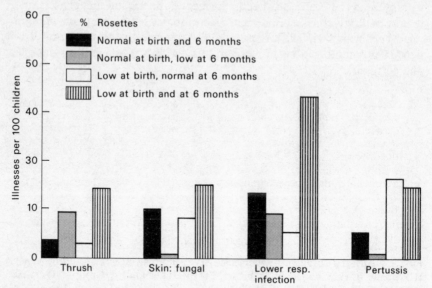

Fig. 2.4. Selected infectious illnesses at 0–6 months and CMI status in Kenyan infants.

immune system[43] compared with the full-weight term infant. Thus, the IUGR infant is placed in double immunological jeopardy with CMI—a precarious situation.

Maternal malnutrition appears to exert its main effects upon the immune system of the developing fetus by interfering with normal growth and development of fetal lymph organs thymus and spleen which in turn decrease T-cell production. Deficiencies of energy, protein, and a variety of specific nutrients limit synthesis of protein and DNA and cell division and growth. The role of stress produced by the conditions in the pregnant women which are associated with fetal malnutrition may result in increased transplacental transfer of cortisol or other hormones. Cortisol in particular is deleterious to lymphoid tissue and T-cell development as seen in acquired postnatal immunodeficiency and malnutrition.[19,40]

With the present level of knowledge, it is clear that any measures that decrease the chances of intrauterine-growth retardation and low birth weight usually help protect the integrity of the developing immune system, particularly the precursors of the T cell—the basic cell in CMI. A comprehensive approach directed at optimal nutrition at the pre-, intra- and post-pregnancy phases of life, the prevention and control of infections, the early detection and vigorous management of diabetes and hypertension, and child spacing are strongly indicated to ensure optimal neonatal and post-neonatal immune function and resistance to infection.

References

1. Jelliffe, D. B. 'Special problems in different groups: pregnant and lactating women'. In *The assessment of the nutritional status of the community*, Monograph Series 53. World Health Organization, Geneva, pp. 210–13 (1966).
2. Tafari, N. 'Low birthweight: an overview'. In *Advances in international maternal and child health, Vol. 1* (ed. D. B. Jelliffe and E. F. P. Jelliffe). Oxford University Press, pp. 105–27 (1981).
3. Susser, M. 'Prenatal nutrition, birthweight, and psychological development: an overview of experiments, quasi-experiments, and natural experiments in the past decade'. *Am. J. clin. Nutr.* Suppl. **34**, 784 (1981).
4. Roeder, L. M. and Chow, B. F. 'Maternal nutrition and its long-term effects in the offspring'. *Am. J. clin. Nutr.* **24**, 812 (1972).
5. Lechtig, A., Habicht, J. P. Delgado, H., Klein, R. E., Yarbrough, C., and Matorell, R. 'Influence of maternal nutrition on birthweight'. *Am. J. clin. Nutr.* **28**, 1223 (1975).
6. Burke, B. S., Beal, V. A., Kirkwood, S. B., and Stuart, H. C. 'The influence of nutrition during pregnancy upon the condition of the infant at birth'. *J. Nutr.* **26**, 569 (1943).
7. Higgins, A. C., Crampton, E. W., and Moxley, J. E. 'Nutrition and the outcome of pregnancy'. In *Endocrinology* (ed. R. O. Scow). Excerpta Medica, New York, pp. 1070–7 (1973).
8. Naeye, R. L., Blanc, W. and Paul, C. 'Effects of maternal malnutrition on the human fetus'. *Pediatrics* **52**, 494 (1973).
9. Fitzhardinge, P. M. and Steven, E. M. 'The small-for-date infant. I. Later growth patterns'. *Pediatrics* **49**, 671 (1972).
10. Chandra, R. K. 'Fetal malnutrition and post-natal immunocompetence'. *Am. J. Dis. Child.* **129**, 450 (1975).
11. Ferguson, A. S., Lawlor, G. J., Neumann, C. G., Stiehm, E. R., and Oh, W. 'Decreased rosette forming lymphocytes in malnutrition and intrauterine growth retardation'. *J. Pediatr.* **85**, 717 (1974).
12. Neumann, C. G., Stiehm, E. R., Swendseid, M., Zahradnick, J. Newton, C., Cherry, J., and Carney, J. 'Longitudinal study of immune function in intrauterine growth retarded infants'. *Fedn. Proc.* **39**, 888 (1980).
13. Singh, M., Manerikar, S., Malaviya, A. N., Premawathi, Gopalan, R., and Kumar, R. 'Immune status of lowbirth weight babies'. *Ind. Pediatr.* **XV**, 563 (1978).
14. Bhaskaram, C., Raghoramulu, N., and Reddy, V. 'Cell-mediated immunity and immunoglobulin levels in light-for-date infants'. *Acta paediatr. scand.* **66**, 617 (1977).
15. Gebhardt, B. M. and Newberne, P. M. 'Nutrition and immunological responsiveness: T- cell function in the offspring of lipotrope and protein deficient rats'. *Immunology* **26**, 89 (1974).
16. Chandra, R. K. and Au, B. 'Single nutrient deficiency and cell-mediated immune responses. I. Zinc'. *Am. J. clin. Nutr.* **33**, 736 (1980).
17. Robson, L. C. and Schwartz, M. R. 'Vitamin B_6 deficiencies and the lymphoid system'. *Cell Immunol.* **16**, 145 (1975).
18. Chandra, R. K. 'Biological Implications'. In *Nutrition, immunity and infection* (ed. R. K. Chandra and P. M. Newberne). Plenum Press, New York, pp. 181–96 (1977).
19. Alleyne, G. A. and Young, V. H. 'Adreno-cortical function in children with severe protein-calorie malnutrition'. *Clin. Sci.* **33**, 189 (1967).

20. Lechtig, A., Yarbrough, C., Delgado, H. Matorell, R., Klein, R., and Behar, M. 'Effect of moderate maternal malnutrition on the placenta'. *Am. J. Obstet. Gynec.* 123 (1975).

21. Mata, L. J. and Villatoro, E. 'Umbilical cord immunoglobulins'. In *Malnutrition and the immune response* (ed. R. M. Suskind). Raven Press. New York, pp. 333–9 (1977).

22. Gross, R. L., Reid, J. V. O., Newberne, P. M. Burgess, B., Marston, R., and Hift, W. 'Depressed cell-mediated immunity in megaloblastic anemia due to folic acid deficiency'. *Am. J. clin. Nutr.* **28**, 225 (1975).

23. Hurley, L. S. 'Trace elements II: manganese and zinc'. In *Developmental nutrition.* Prentice-Hall, New Jersey, p. 99 (1980).

24. Miller, M. E. 'The immunodeficiencies of immaturity'. In *Immunologic disorders in infants and children* (ed. E. R. Stiehm and V. Fulginiti) 2nd edn. Saunders, Philadelphia, pp. 219–38 (1980).

25. Arenson, E. B., Epstein, M. B., and Seeger, R. C. 'Monocyte Subsets in neonates and children'. In *Host defenses in the fetus and neonate, Pediatrics* Suppl. **64**, 740 (1979).

26. Blaise, M., Poplack, D. G., and Muchmore, A. 'The mononuclear phagocyte system role in expression of immunocompetence in neonatal and adult life'. *Ibid*, 829 (1979).

27. Quie, P. G. and Mills, E. 'Bacteriocidal and metabolic function of polymorpho-nuclear leukocytes'. *Ibid*, 719 (1979).

28. Miller, M. E. 'Phagocyte function in the neonate: selected aspects'. *Ibid*, 709 (1979).

29. Johnson, R. B., Attenburger, M. S., Atkinson, A. W., and Curry, R. 'Complement in the newborn infant'. *Ibid*, 781 (1979).

30. Hayward, A. R. and Lydyard, P. M. 'B-Cell function in the newborn'. *Ibid*, 758 (1979).

31. Uhr, J. W., Dancis, J., and Neumann, C. G. 'Delayed-type hypersensitivity in premature neonatal humans'. *Nature* **187**, 1130 (1960).

32. Stiehm, E. R., Winter, H. S., and Bryson, Y. S. 'Cellular (T-cell) immunity in the human newborn'. In *Host defenses in the fetus and neonate, Pediatrics* Suppl. **64**, 740 (1979).

33. Jelliffe, D. B. and Jelliffe, E. F. P. 'Protection and hazards'. In *Human milk in the modern world.* Oxford University Press, London, pp. 84–112 (1978).

34. Nutrition Committee of the Canadian Paediatric Society and The Committee on Nutrition of the American Academy of Pediatrics. 'Breast Feeding'. *Pediatrics* **62**, 591 (1978).

35. Hanson, L. A. and Winberg, J. 'Breast milk and defense against infection in the newborn'. *Arch. Dis. Child.* **47**, 848 (1972).

36. Chandra, R. K. 'Immunocompetence in Undernutrition'. In *Nutrition, immunity, and infection* (ed. R. K. Chandra and P. M. Newberne). Plenum Press, New York, pp. 67–120 (1977).

37. Collaborative study of Ministry of Health Kenya and WHO. 'Measles immunity in the first year after birth and the optimum age for vaccination in Kenyan children'. *Bull. WHO* **55**, 21 (1977).

38. Jagadeesan, V. and Reddy, V. 'Serum complement and lysozyme levels in light-for-date infants'. *Acta paediatr. scand.* **67**, 237 (1978).

39. Manerikaris, S., Malaviya, A. N., and Singh, M. B. 'Immune status and BCG vaccination in newborns with intrauterine growth retardation'. *Clin. exp. Immunol.* **26**, 173 (1976).

40. Neumann, C. G., Lawlor, G. J., Stiehm, E. R., Swendseid, M. E., Newton, C., Herbert, J. Ammann, A., and Jacob, M. 'Immunologic responses in malnourished children'. *Am. J. clin. Nutr.* **28**, 89 (1975).
41. Ferguson A. C. 'Prolonged impairment of cellular immunity in children with intrauterine growth retardation'. *J. Pediatr.* **93**, 52 (1978).
42. Metcoff, J. 'Maternal leukocyte metabolism in fetal malnutrition'. *Adv. exp. Med. Biol.* **49**, 73 (1974).
43. Neumann, C. G., Stiehm, E. R., and Cherry, J. 'Immune function and infection in intrauterine malnourished infants'. In *Proceedings: infections in the immunocompromised host-pathogenesis, prevention and therapy* (ed. J. Verhoef, P. K. Peterson and P. G. Quie). Elsevier, Amsterdam (1980).

3 Integrating traditional and modern health systems

R. H. O. BANNERMAN

Introduction

Traditional medicine is a term loosely used to distinguish ancient health-care systems from official modern scientific medicine or allopathy, and includes medical beliefs, practices, and materials which existed before the application of science to health matters. Other terms often used as synonyms are indigenous, unorthodox, alternative, folk, ethno, unofficial and fringe medicine and healing. The term is unsatisfactory because it implies broadly that there is some body of principle, knowledge, and skills common to all the varieties of traditional medicine; and because it does not distinguish between the all-embracing and complex systems of health care such as Ayurveda on the one hand, and simple home remedies on the other.

Traditional medicine has been practised to some degree in all cultures and other terms based on culture include 'African, Asian or Chinese medicine'. No health service starts in a vacuum—the healers, the people, and their ailments are already in existence and every society has responded to the challenge of sickness with its own health beliefs and practices.[1]

Traditional societies regard health as a state of balance or equilibrium both internally and externally. This equilibrium is based on variations of humoral substances and forces including balances of the opposing qualities of hot and cold, and wet and dry; portraying in effect the Chinese principle of yin and yang.

Traditional practitioners in many parts of the world define life 'as the union of body, senses, mind, and soul', and describe positive health as 'the blending of physical, mental, social, moral, and spiritual welfare'.[2] The moral and spiritual aspects of life are here stressed, thus giving new dimensions to the system of health care by which man maintains his health. The authority of a traditional healer may be derived solely from his recogition in the community in which he lives as competent to provide health care; or from his qualification in an elaborate formal system, such as Ayurveda. His practice generally follows the social, cultural, and religious background of the community and he will use vegetable, animal, and mineral substances, and methods and ceremonies handed down to him by tradition, to intervene in physical, mental, and social conditions related to health and sickness.

There are those who take a hostile attitude towards traditional medicine and those who, without reservation, accept all things handed down by tradition. Both of these attitudes are indeed wrong. The more realistic tendency which is emerging is discriminating and discards the crude and harmful practices while retaining the refined and useful methods for further development and application. The same applies to the clinical approach to modern scientific medicine, particularly in the developing countries, where not infrequently technical bias makes it incomplete and indeed potentially harmful because it pays little or no heed to socio-cultural factors, and is focused mainly on laboratory diagnosis.

The belief that illness arises from supernatural causes and indicates the displeasure of ancestral gods and evil spirits, or is the effect of black magic, is still held, by many communities in Africa, Asia, and the developing world; and to some extent this is true also of the industrialized countries. It is therefore wrong to attribute magical, irrational, and superstitious ideas to any group of countries or levels of industrial or educational development. The evidence is that the two approaches to health care are complementary, and that with the swing of the pendulum greater attention should be paid to the traditional practices which bring comfort to very large numbers of people everywhere.

Diseases have been classified into three broad categories—those curable by indigenous or traditional medicine, those curable by modern or western medicine, and those that are self-limiting and not affected by either indigenous or modern medicine. Health behaviour remains a rational response to the perceived causes of illness and tends to be adaptive and to promote the survival and increase of the numbers in each society.

Health care delivery[4]

Gaps in our knowledge and lack of resources are not always the main reasons for deficiencies in the health care delivery system. Unfortunately, limited resources are often not utilized to the full, and this defect often relates to existing cultural barriers in the communities concerned.

National health planning is defective in many countries and the baseline national health statistics and other relevant information essential for realistic health planning are often not available. For example, in countries where less than 20 per cent of the babies are delivered by health professionals, pertinent questions should be asked by the health administrators on the fate of the other 80 per cent of the total births in the country. Who are responsible for the care of these mothers and babies? The obvious answer is 'the traditional midwives'! How many are they? What roles and functions do they have and what training? These are but a few of the obvious questions that should engage our attention. Besides, the small number of health professionals, relatively ineffective in the rural areas where their services are most needed, tend to be concentrated in the urban areas, where the ratio is about one doctor per 2–3 thousand people; yet in the rural areas, where about 80 per cent of the population lives, the ratio is about one doctor per 50 thousand people. As populations in the developing countries grow and doctors emigrate, the imbalance becomes even worse in many countries. Healers and traditional midwives, on the other hand, practise in almost every village in some countries and so are certainly more accessible to most people than either doctors, nurse/midwives, or hospitals.

Health professionals are in the main unwilling to delegate responsibility and to allow non-professional health workers to take over parts of their professional roles. Often these professionals responsible for primary health care are inadequately trained for the work which needs to be done. The end-result is that the return for the resources and human effort is poor and well below expectation; but the professional health worker has invariably been trained in a scientifically oriented medical centre whilst endeavouring to work in communities that are essentially rural, traditional in outlook, and with very different cultural backgrounds.

Traditional medicine is an established part of culture, though in some countries the systems of care and prevention may not be as well developed as in China and other Asian countries. Some countries have retrained indigenous traditional healers for work in the official health care systems and have established departments for traditional medicine in the ministries of health and the universities. For example, research in African traditional medicine and pharmacopoeia is being encouraged, particularly by the scientific branch of the Organization of African Unity, and associations of healers are being established with Governments' support.

Studies have been started to determine the possibility of local manufacture of pharmaceutical products, including traditional African medicines and medicinal products. Such activities have long been established in China and several other countries in Asia and Latin America.

The various practitioners have been observed to 'borrow' freely from one another whenever this appeared advantageous to the health situation. As this 'borrowing' became more widely accepted, planning for integration was achieved through carefully structured training programmes, integrated multidisciplinary research, the teaching of basic and fundamental principles, investigation of medicinal plants and herbal remedies, and the testing of hypotheses as a basis for further research.

In the USA the process of connecting Amerindian medicine with Government medicine began over 200 years ago. The early European settlers depended on traditional methods and healers and when Amerindians had health problems not amenable to traditional methods, they equally resorted to western medicine. The two systems therefore functioned side-by-side and the dual system persists today. Most of the healers function on a full-time basis, but a few work in other capacities such as business, management, and Government services. Remuneration for services varies; some charge flat rates, others accept gifts, and some will not accept anything.

The health hierarchy in traditional societies does not operate rigidly.

People sometimes try several treatments at once. They move back and forth from herbalist to shaman, and to modern drugs and modern hospitals. This flexibility allows for effective and timely intervention by professional health workers if they could only appreciate what their appropriate role might be within the structure of their community. Education and the mass media are producing rapid changes in these communities and the value of modern medicine is now known everywhere, but its acceptance varies from country to country. The urban élite tend to use modern medicine more than the urban poor and rural communities. In many areas both the rural and urban poor have realized the efficacy of modern drugs in the control of infectious diseases, and the virtues of surgery.

The most susceptible groups for the introduction of modern medicine are the urban poor—those immigrants from the rural villages who have lost their traditional ties and built-in social security systems based on the extended family, and are anxious to become modern.

Large city hospitals are usually similar in construction and in the services administered; however, the small rural health posts should tend to reflect more of the cultural milieu within which they operate in order to gain acceptance with the community.

Most traditional societies are aware of modern health facilities and will accept such resources if offered with sympathy and understanding, and with due regard for the health care already prevailing locally.

The felt and real health needs of rural communities vary throughout the world and can be determined effectively only by on-the-spot study and analysis. Without such basic information, it would be difficult to plan modern health services realistically. Health care programmes designed for rural England will not be automatically suitable for rural Ghana.

The health authorities that wish to train and encourage traditional midwives to participate actively in national maternal and child health programmes should first determine what these midwives are doing in their respective communities, since programmes designed without sufficient account of prevailing cultural practices often do not succeed.

Such training programmes have to be carefully structured, with the inclusion of some senior traditional midwives in the planning stage. The trainers themselves should be selected on account of their knowledge of local customs and health problems, teaching, and service experience, and should be acceptable to the traditional midwives. Before such assignment, they would require some orientation in the training of illiterate and semi-literate adult persons with no previous formal education.

Traditional practices in maternal and child health services

About 80 per cent of babies born in the developing countries are delivered by traditional midwives or traditional birth attendants.[5] Other synonyms used for this handy and ubiquitous health worker include lay, native, empirical, indigenous, dai (in India), and other local terms. The term 'traditional birth attendant' or TBA was introduced by the World Health Organization and is defined as 'a person who assists the mother at childbirth and who initially acquired her skills delivering babies by herself or by working with other traditional birth attendants'. The term 'midwife' is therefore used to refer to a person with formal health-sciences education and officially registered or licensed. Several health administrators and others have commented that the term 'traditional birth attendant' is too narrow and does not reflect the functions and role of the local midwife in the community. The term traditional midwife will therefore be used in this text.

The practice of midwifery—helping women to give birth—has always existed. Throughout the Greek and Roman periods of European history and during the great Islamic Empire period, midwives were known to have practised alongside university-trained physicians without conflict. Normal midwifery remained the responsibility of the midwife whilst physicians attended to abnormal and emergency cases.

In Western Europe, until about the thirteenth century, women practised actively as traditional healers and attended to the health needs of their communities.

Late in the medieval era however, the Church began to assume greater power in Europe and the clergy tried to bring all ritual and religious functions under its control. Since midwives depended on rites and potions to protect women in labour, they became victims of the clergy's efforts to root out these old religious practices. Midwives were often accused of witchcraft and of causing congenital defects and making women sterile, and many such midwives were executed. During this same period, the development of professional bodies amongst physicians and surgeons also hindered the training and licensing of midwives.

Traditional midwives in other parts of the world, however, managed better than those in Britain. In France and Germany midwifery practice was regulated early in the fifteenth century by law, and training courses were established in medical schools and hospitals in the early eighteenth century.[6]

Today there are many different categories of midwife, ranging from the uneducated traditional midwife or so-called traditional birth attendant to the highly trained nurses and midwives who have graduated with masters and doctorate degrees from university teaching hospitals.

Just as the developed countries faced difficulties accommodating midwives in the health professions, so also the developing countries in Africa, Asia, and Latin America face similar problems in collaborating with their traditional midwives to help improve national health services.

The traditional midwife's profile and work

Traditional societies view childbirth as a component of the traditional health system and with social, cultural, and moral significance. Though many communities emphasize the normal physiological aspects of pregnancy and childbirth, some consider the event as an illness and with considerable anxiety. Various precautions are therefore taken and rituals are performed for the protection of both mother and child. Similar rituals and ceremonies are also performed at the menarche to mark the advent of womanhood. Pregnancy and childbirth produce temporary states of imbalance and much care is therefore taken to restore the equilibrium, usually through prescribed diet and restriction of activities. Postpartum mothers, for example, are considered to be in a cold state on account of bloodloss during delivery—blood being considered 'hot'. Cold foods are therefore prohibited whilst hot baths and teas brewed with herbs that are reputed to have hot qualities are encouraged.[7]

Blood is considered unclean and menstruating or newly delivered women are equally regarded as contaminated and should therefore be avoided. since blood is associated with childbirth many traditional midwives tend to wear dirty apparel for the delivery, but with tactful orientation and training it has been possible to wean several traditional midwives from this unhygienic practice and they now wear clean clothes and aprons and work in surgically clean conditions.

In most developing countries, traditional midwives are middle-aged or elderly, illiterate, married, or widowed with several children, and practise midwifery on a part-time basis. Age and experience command respect in traditional societies and traditional midwives generally enjoy high status, except in India where midwifery is delegated to women of low caste. In some countries, a number of traditional midwives are men—usually herbalists who practise midwifery as a specialty. In a recent community health survey in Southern Ghana (Danfa), over 50 per cent of the traditional midwives were found to be male.[8]

Traditional midwives acquire their skills and knowledge through several years of apprenticeship to an older established midwife who may be a parent, close relative, or family friend. They learn informally by observation and practice, and often inherit their clientele from their mentors and other older midwives. The practice may thus be restricted to a group of extended families. Midwifery is often a part-time job and the midwife may engage in retail trading, agriculture, or other activities. The practice also varies in intensity—some may have as many as ten or more deliveries a month whilst others conduct only one or two births throughout the year.

The traditional midwife develops a personal, informal relationship with her patients and offers much psychological support in contrast to the professional midwife who in such situations is authoritative, impersonal and formal. The traditional midwife knows the client and her extended family, generates confidence, and shows interest in all aspects of the patient's life.[9]

Their functions are indeed diverse and some could be described as 'traditional healers' who specialize in midwifery and reproductive ailments. They not only deliver babies but also assist their patients during the prenatal and postnatal periods, and give advice on child care, infant feeding, lactation, and weaning. Some offer treatment for infertility, advise on contraception and even perform abortions and ritualistic surgical procedures such as female circumcision.

The pregnant woman will visit the traditional midwife early in pregnancy and that may be any time between the third and fifth month of pregnancy. The traditional midwife might determine the expected date of delivery, observe her general physic, palpate the abdomen to determine the position and size of the fetus, and check the mucous membranes for pallor. She is then given advice on general life style during pregnancy, clothing, household and other activities which are usually restricted, sexual relations with her husband, and taboos and rituals to be observed. The traditional midwife would know the woman's family and cultural background well and mutual confidence is established at the first visit. Where the midwife is also a family friend, she might make the first visit to the patient's home, confirm the diagnosis of pregnancy, and conduct the first examination there.

Some mothers restrict food intake because of morning sickness and fear of a large fetus. The inclusion of raw liver in a mixed diet is often encouraged, especially where there is pallor of the mucous membranes. In certain African communitites, chicken is favoured but fish is restricted in the belief that it could cause hardening of fetal bones and thereby induce dystocia in labour.

The prenatal massage has high priority in many developing countries. It is administered sometimes with medicated oils which are reputed to make the delivery easier and to prevent pregnancy and postpartum pains.

Visits are made at fairly regular monthly intervals and some midwives examine the urine for normal colour and smell.

During palpation, any abnormal fetal lie, including the breech position, is corrected to a cephalic presentation by external version. The experienced traditional midwife is usually adept at these manoeuvres and will listen for any changes in the fetal heart rate by applying her ear directly on to the mother's abdomen. Some now use a clay or wooden fetal stethoscope.

In some African communitites the traditional midwives will dilate the vagina manually using medicated oils during the last few weeks of pregnancy to facilitate delivery and to prevent perineal lacerations.

The massage is considered essential in late pregnancy and many women therefore prefer to transfer from the hospital clinic to the traditional midwife to ensure having this massage.

Medication in the form of vegetable tampons inserted into the vagina late in pregnancy and early labour is not uncommon, and may be a source of genital infection with serious consequences.

The majority of deliveries in rural areas are conducted in the patient's home. Recently, however, some midwives—especially the trained ones— have prepared a delivery room in their own homes where they receive their patients for delivery. The room has very little furniture, a low stool on which the midwife sits, a small table for the delivery bag containing a pair of scissors, cord dressing bowl or dish, rubber catheter, etc. The mother is taken home within 24 hours after delivery. The advantage of this arrangement is that the midwife is able to attend to more than one labour patient at any given time and the delivery is conducted in relatively clean and hygienic surroundings. Besides, should emergency help be summoned, the midwive's address is well known to the district hospital authorities.

In most traditional societies childbirth is more of a social event than a medical one and women always expect a normal delivery, regarding the event as an achievement and proof of womanhood. A woman that has an assisted delivery, particularly in the form of a caesarean section, is considered a failure and may be taunted with it. Health professionals should therefore explain very carefully to both the patient's family and the traditional midwife the reasons for surgical interventions during delivery and that such actions are in the best interests of both mother and child. Otherwise, with a subsequent pregnancy the mother may well rigorously avoid hospital delivery, with disastrous results.

During labour, the traditional midwife gives the mother all psychological support and advises her regarding what to expect in terms of labour pains and when to bear down to effect delivery. Shouting or crying is disapproved. The squatting, kneeling, or sitting positions are usually preferred and in some communities assistants support the woman in a semi-reclining position, sitting on a mat or floor mattress. The left lateral or supine positions usually adopted in western societies is alien to traditional societies and, although preferred by the obstetrician and western-trained midwife, are more difficult for the mothers. These positions with associated isolation may be another reason for mothers evading hospital delivery.

Some traditional midwives remain relatively inactive and allow labour to progress naturally.[7] Others, however, sometimes administer herbal infusions with oxytocic components and even massage or exert physical pressure on the uterine fundus to aid delivery. Such activities are known to cause tonic uterine contractions and rupture. However, these are gradually being discontinued through training and orientation programmes for traditional midwives.

One or more persons, usually an elderly female relative, attends the birth in addition to the midwife. Surgical interference is rare but traditional midwives are known to perform episiotomies, expecially in women who have undergone circumcision with resultant scarring and narrowing of the vagina. Immediately after the birth of the baby, the cord is cut with some household implement, razor blade, or scissors which may not be necessarily clean or sterile, and dressed with salt, clay, palm oil, wood ashes, spider webs, herbal preparations or dried cow dung, depending on the locality. Experimental evidence from Central America has shown that spider web has antiseptic properties.[10] Manual removal of the placenta is rarely done. The use of dried cow dung for dressing the umbilical cord stump is particularly dangerous and undesirable. This is done on account of the desiccating properties of dried cow dung and facilitates early separation of the cord. Simultaneously, however, this act introduces tetanus bacilli, which cause neonatal tetanus with high mortality and morbidity rates. Health education and appropriate in-service training are currently being used in many countries to improve midwifery practice amongst traditional midwives and to eliminate such harmful practices. In cases of difficulty, the traditional midwife may perform rituals, say prayers, and listen to confession.

After delivery the midwife and the grandmother or other elderly relative stay with the mother to give all necessary care and support, including staving off evil spirits. The placenta is ceremoniously buried in the compound in the belief that should the child reach adulthood and sojourn to some distant land, the placenta will ultimately draw him back to his place of birth. On the seventh day, or a month or two after delivery, and depending on the community's traditions, a naming ceremony is performed in the presence of the extended family and friends.

Many traditional midwives take active steps to hasten the recovery of the mother after delivery, and use a wide range of traditional measures such as massage, ceremonial and sitz baths, binding and herbal medications to stop any heavy postpartum bleeding. Some midwives stitch perineal wounds and dress them with alcohol, salt and various herbs to prevent infection.

Traditional child care

This is generally based on good sense and long experience in the community. The traditional midwife continues to take care of both mother and child for several weeks after delivery. Breast feeding is encouraged for at least 6 months and during this period lactation is ensured through breast massage, liberal diets, and lactogenic herbal infusions. Breast feeding may be continued for some 2 years, partly as a means of staving off pregnancy. The mother is confined to the house for 1–4 months, during which period her activities are restricted. Should the mother be severely ill or lactations cease, a wet-nurse would be engaged from within the community to breast feed the baby. Early weaning is discouraged, except where an unexpected pregnancy supervenes.

Weaning practices vary and in some communities forced feeding is undertaken—usually with the child in the horizontal position, which makes swallowing difficult and could cause respiratory complications through inhalation. Some of these children are put on corn broth with no protein or other supplement, thus exposing them to the malnutrition syndrome—kwashiorkor.

Other harmful practices include the cauterization of the child's gums in the erroneous belief that this aids dentition, likewise cauterization of the child's back to stop diarrhoea, and excision of the uvula in cases of whooping cough. Ordinary water is discouraged in certain communities for fear of gastro-enteritis through contamination, and non-fermented local beer prepared by boiling for some 3 days is given instead. The brew is essentially non-alcoholic and until safe water is assured this is no doubt a practice to be encouraged.

Fertility regulation

Traditional communities consult healers and traditional midwives on the treatment of infertility, prevention of unwanted pregnancy, and procurement of abortion. Advice is also sought on unsatisfactory marital relations and gynaecological problems such as low backache, lower abdominal pains, dyspareunia, menstrual irregularity, etc. Infertility is a matter of grave concern, since the barren woman is often ridiculed and even divorced; needless to say that in such communities the blame for infertility is always totally on the woman. Various rituals are performed, including a visit to a cemetary by the infertile moslem woman who voids urine on the grave of a Christian in the hope that she becomes pregnant soon afterwards.[11] The placenta too may be used for such purposes and in certain African communities it is handed to an infertile woman, anxious to procreate, for ceremonial burial. An Ashanti fertility doll may also be hung in the living room. Various herbal medicines, vaginal suppositories, warm baths, abstinence then intercourse on prescribed days are also used. When therapeutic measures appear unsuccessful, it is not unusual for either the infertile wife herself, or the husband's mother, to choose a younger wife for the husband.

The most common traditional methods recommended by midwives are withdrawal, abstinence, and abortion. Abstinence is common mainly in polygamous societies and is often backed by taboos and local customs. The woman may be obliged to sojourn to her parents for about 2 years and until the child can walk. Whilst she is lactating and breast feeding sexual intercourse is prohibited.

Some traditional midwives perform abortions and among the methods used are herbal preparations taken in the form of teas or herbal tampons inserted into the vagina.

Fertility regulation has become a major concern of environmentalists, politicians, and Governments in both the industrialized and third-world countries; and although several approaches are currently being used throughout the world for fertility regulation, none appear to be entirely satisfactory, and sociological, cultural, clincial, practical, and economic constraints have been widely reported. However, a particularly attractive approach for the developing countries would be through the utilization of plants proved effective in traditional medicine.

Several plant constituents have been shown to elicit one or more types of antigonadotrophic activity, but the usefulness of such agents for human fertility regulation is said not to be practical. However, about 300 plant species have been reported to exhibit a variety of fertility-regulating effects.[12]

Perhaps the most intriguing of the fertility-regulating agents in the plant kingdom is a simple aromatic compound, *m*-xylohydroquinone (2,6-dimethylhydroquinone), first isolated from the common garden pea (*Pisum sativum* L.) in 1952. Interestingly, peas constitute the major protein staple of the diet in Tibet and the population in that country has remained essentially static for the past 200 years.

Among the various precursors of cortisone and sex hormones are the yams, or Dioscoreae, first identified in Mexico but common also in tropical Africa, China, and India.

The tubers of these yams are rich in carbohydrate and are used widely for food in tropical countries. Their hormone components form the basis for making the contraceptive pill. Yams are grown mainly in China but grow freely in Asia and Africa. Sisal is grown in tropical America and Africa and in this instance the leaf waste stripped during removal of the fibre is fermented to produce the requisite sex-hormone precursors.

Regarding male contraceptive agents, the Chinese report that the cotton-seed oil extract Gossypol has been found to inhibit the motility of spermatozoa to over 90 per cent within 1 week after the administration of a therapeutic dose. The effect is said to last from 3 to 4 weeks.[4]

Female circumcision in children[4,11]

A highly undesirable traditional practice affecting the health of children is female circumcision. This is a ritualistic operation performed on children between 1 and 15 years of age, mainly by traditional midwives. A few such operations are, however, undertaken by doctors, nurses, and midwives. Three types are described:

1 *Circumcision*—excision of clitoral prepuce. Sometimes referred to as Sunna. It is occasionally performed in frigid women to facilitate induction of orgasm but has been largely discontinued on account of unsatisfactory functional results.

2 *Excision*—removal of clitoris together with adjacent labia minora.

3 *Infibulation* or *pharaonic circumcision*—removal of clitoris, adjacent labia minora and anterior two-thirds and often the whole of median part of the labia majora is also removed. This type may produce serious immediate and remote or long-term complications. The ritual is performed in certain African countries, parts of Asia, Australia, and the Americas.

Various reasons are given for this operation and they include personal hygiene, elimination of promiscuity, safeguarding virginity until marriage, the prevention of sexual assaults on young girls and also to make them less attractive for the slave market. In several countries legislation has been enacted to make the operation illegal, but some families still demand it for their daughters, and husbands also expect their young brides to have been circumcised. The enforcement of such laws has therefore been difficult.

Complications such as shock, haemorrhage, pain, chronic infection, vaginal fibrosis, dyspareunia, and even vesico-vaginal and recto-vaginal fistulae, and dystocia in labour are all well recognized.

The world press has recently focused attention on this problem, and health education of the public and communities concerned is being actively pursued, with good results.

There is need for a clearly defined national policy to control and finally abolish female circumcision. Control may sometimes mean transferring the circumcision ritual from the unhygienic back-streets and village alleys to the more favourable hospital setting—a choice between the lesser of two evils. General education of the public and appropriate training for traditional birth attendants and healers could do much to eliminate this practice, and here religious and community leaders in the countries concerned have an important role to play.

Integration[11]

Many public health administrators now concede that healers and traditional midwives have a role to play in formal health services, and certain countries now consider the concept of integration a reality that could be achieved in the foreseeable future. In China and India, traditional systems of medicine have already been recognized, legalized, and well developed as separate systems in their own rights.

The integration of the various systems of health care offers several benefits to each system and helps to improve the general health care, especially of the disadvantaged and deprived rural populations. It enhances the quality of the various practitioners and promotes the dissemination of knowledge in regard to primary health care. There are, of course, problems relating to integration, such as fundamental differences between the concepts of life, health, and disease underlying the philosophies on which the various medical systems are founded. One major constraint is nomenclature of diseases, definitions, and diagnosis. For instance, healers often have only one generic term for tumours and this embraces both innocent and malignant neoplasia. There is need for more effective communication, a clearer definition of terms and valid methods for evaluation of therepeutic methods.

Ideally, to meet the compelling health needs of the developing countries, health professionals should work in collaboration with a broad range of indigenous practitioners. These healers could learn how to use many modern drugs and techniques safely; in turn, scientifically trained practitioners could learn both how to apply culturally accepted indigenous practices or adapt them in such a way as to make them safer, and how to introduce modern practices in such a way as to make them culturally more acceptable. In practice, however, hostility or at the very least poor communication between the educated health professionals—physicians, nurses, and midwives—and the healers and traditional midwives is often cited as a major problem in such community-health endeavours.

Today, a dual system of medicine exists in many developing countries—a modern/scientific one and a traditional one. The modern health sector consists mainly of university-educated physicians, practising in western-European type of private clinics, urban hospitals, and research centres.

The traditional health sector, on the other hand, consists of a great variety of indigenous practitioners, including traditional midwives, who live close to their clients, dispense herbs, potions, massage, and other familiar remedies, and also provide conventional wisdom and personal services for people who speak the same language and come from the same background. For most villagers and many city dwellers in developing countries, going to a traditional practitioner is more comfortable, more convenient, and usually less expensive than going to a physician or hospital. Unless patients are sure that only a physician and modern technology can take care of their problem they are more likely to try the traditional practitioner first.

Many countries and international agencies have undertaken programmes to train and improve the skills of traditional midwives. In 1952, the United Nations Children's Fund (UNICEF) began to organize training programmes and to provide simple midwifery kits to traditional midwives that completed their training sessions. During the 1960s the International Planned Parenthood Federation began to study the use of traditional midwives in family planning programmes; the Governments of Pakistan and India initiated national family planning programmes in the late 1960s which relied heavily on the nations' dais.

Traditional midwives are envisaged by several health administrators as an important health manpower potential for extending maternal and child health care to the vast populations in the rural area. However, a crucial factor contributing to much of the failures in these programmes is the poor information relating to the traditional midwives' beliefs, role, and functions in the community. The cultural aspects of the whole process of pregnancy, childbirth, and parenthood are often ignored by the physician and professional midwife, thus reinforcing the cultural gap between the traditional and the modern.

Physicians practising in traditional communities often have difficulty in communicating with patients because they are often ignorant of traditional medicine and fail to understand its vocabulary and rationale. Recent evidence shows, however, that doctors and other health professionals now have better appreciation of their patients' views on disease aetiology and are able to communicate with them more effectively.

Where age confers respect, the imposition of a young, unmarried professional midwife as supervisor to traditional midwives can create considerable conflict and can undermine the value put on age in traditional communitites. The choice of a midwifery teacher or supervisor should therefore be made with due consideration.

Despite the high status in the community held by these traditional midwives, professional health personnel and educated persons tend to treat them disparagingly and to regard them as superstitious, ignorant, and dangerous. Such notions hinder effective collaboration and should be discarded.

In traditional societies, the patient's mother, elderly aunt, or grandmother retains much of the decision-making power though she often consults with the traditional midwife; whereas in western societies the obstetrician maintains full authority.

The outcome of the birth is believed to be up to God, the traditional midwife acting only as his agent. Accountability is therefore to God. However, she is also accountable to her community and the size of her practice will be influenced by the number of successful deliveries.

The Chinese example

The People's Republic of China provides a very good example of what health benefits can be obtained in a relatively short period of time through the integration of the traditional and modern systems. Since 1949, the then existing meagre health services, which were insufficient even to serve a small fraction of China's vast population, have been increased by the integration of traditional Chinese medical techniques and practices with western medicine, and by the training of primary health care workers in adequate numbers, popularly known as 'barefoot doctors', who have the enthusiasm and skills to deploy such an integrated form of medicine through the whole countryside.[4]

Research institutions have been established and top-level scientific research and studies are being conducted into techniques, such as acupuncture for anaesthesia and treatment.

Acupuncture

'That a needle stuck into one's foot should improve the functioning of one's liver is obviously incredible. It can't be believed because, in terms of currently accepted physiological theory "it makes no sense". Within our system of explanation there is no reason why the needle-prick should be followed by an improvement of liver function. Therefore, we say, it can't happen. The only trouble with this argument is that, as a matter of empirical fact, it does happen.'[14]

The above statement by Aldous Huxley ably summarizes the current situation regarding acupuncture and underscores the fact that we still know relatively little about nature and certainly do not know everything about the structure and functioning of the human body.

Acupuncture has been applied as a therapeutic medical technique in China since at least 2000 years ago, when stone knives and other sharp instruments were used. The term itself is derived from the Latin words *acus*—needle—and *punctura*—puncture.

Until about 50 years ago, the practice of this technique was still confined to those to whom it was handed down from one generation to the other, together with the golden needles which were then in use. Thin filiform needles are inserted into various parts of the body to treat a variety of diseases comprising allergic reactions, bacterial infections, and degenerative disorders. Since 1955, acupuncture has been used successfully as analgesia for both major and minor surgical procedures. Needles are typically left in position from 15 to 30 minutes during treatment, and for much longer periods during surgical operations. They are manipulated in twirling or push/pull movements, or activated by pulsed electrical stimulation. An acupuncture effect can also be obtained by deep finger pressure, so-called acupressure. More recently, stimulation of the recognized acupuncture 'points' have included the use of ultrasound and lasers. These require rather expensive equipment but the risks associated with the insertion of needles into the body are thereby eliminated.

This medical procedure of acupuncture is today being accorded greater attention in several Asian and European countries and also by the World Health Organization under its Traditional Medicine Programme. Acupuncture is an important therapeutic method within the Chinese traditional system of medical practice and is often used in combination with other therapeutic measures and may also be used as a diagnostic aid, for example in conjunction with fluoroscopy in gastrointestinal diseases. Acupuncture is not a panacea for all ills, but the available evidence demands that it be taken seriously as an appropriate health technology of considerable value. Some 200 diseases are thus being treated by such methods. The main attraction is that it has relatively simple techniques and can be readily adapted for application at primary health care level in many Third-World countries.

To understand the mode of action and further develop this ancient technique for wider application, in view of its simplicity and proved therapeutic efficacy, research is being carried out into the mechanism of acupuncture and acupuncture analgesia.[15]

Primary Health Care

The optimum realization of the integration of traditional and modern health care systems is at the level of primary health care, and such health care is the main vehicle through which an acceptable level of health and near total coverage can be achieved in the foreseeable future. The strategies to be adopted will, however, vary from country to country and would depend *inter alia* on the political, social, cultural, and economic patterns.[16]

The developed countries are in a more favourable position as on the whole their facilities and health manpower are both reasonably adequate. What is now required in some developed countries is a redistribution of functions and responsibilitites amongst health personnel and more efficient restructuring to reduce costs and increase productivity. In many countries the soaring costs of health services have become extremely burdensome to the exchequer, and the rationalization of self-care and traditional practices could do much to alleviate some of this burden.

The developing countries, however, pose a picture of want and deprivation with inadequate resources, a dearth of professional health manpower, and with no definite hope of amelioration in the foreseeable future, except through the adoption of unorthodox measures such as the exploitation of useful traditional health practices, including the wider use of locally produced herbal medicines and the incorporation of traditional practitioners into the health team.

In view of the pressing needs of the developing countries, this section will be devoted to those countries.

There is widespread disenchantment with health care in many of these countries and for various reasons which are common to nearly all of them. Health resources tend to be concentrated in urban areas which accommodate only about 20 per cent of the total population. These facilities are so expensive that only the élite and opulent citizens in the city can afford to avail themselves of the specialist services utilizing such costly technology. In the final analysis only a very small proportion of the total population have ready access to such facilities. Meanwhie the so-called orthodox and conventional health care services devised for Third-World populations remain culturally unacceptable and economically unobtainable. The disparity between high costs and low returns in health care has become apparent in both the developed and developing countries, and the economics of health care systems have become a major political issue in many countries.[17] In several developing countries, whilst about 7 per cent of the national budget is allocated to the health services, some 30 per cent of that amount is absorbed by the drug bill alone; and all the drugs for such relatively poor countries are imported with hard currency.

A number of these countries in Africa, Asia, and Latin America are therefore exploring the possibility of developing their well-known and tested herbal medicines for use in primary health care centres. These medicinal plants are generally locally available and relatively cheap and there is every virtue in exploiting such local and traditional remedies when they have been tested and proved to be non-toxic, safe, inexpensive, and culturally acceptable to the community.

The high cost of primary health care has become such a major constraint that what is regarded as basic in health services in the developed countries is considered a specialized service in a developing country.

In many of these countries primary health care devolves on the healer, herbalist, traditional midwife, and other traditional practitioners. They are the health workers that offer services to the disadvantaged groups that total about 80 per cent of the world's population and that have no access to any permanent form of health care.[18] Traditional medicine therefore has a major role in primary health care in terms of numbers of people served by that health care system throughout the world and whatever its defects. As already stated, the traditional practitioners are the true community health workers in their society. They invariably have the confidence of the people, and whatever their level of skills, it is essential that they understand the health needs of their community. There are, however, certain major constraints on the use of traditional medicine in primary health care. these are not unsurmountable and with good will on all sides they can be overcome.

Categories of traditional practitioners

There are four main categories of traditional practitioners. The first are those who have received a fully integrated training in modern and traditional systems of medicine such as in Ayurvedic, Unani, and Chinese medicine. The second are those trained mainly in traditional medicine but who also have elementary knowledge of modern medicine. Such health workers practise mainly in smaller rural communities, using traditional medicine most of the time and modern allopathic drugs in emergencies. The third group are the traditional practitioners without formal training but who have obtained diplomas in some particular traditional system such as Ayurveda after correspondence courses and examinations. They practise only traditional medicine. The fourth group comprises those without either institutional training or qualifications and who practise traditional medicine after several years' apprenticeship with an established traditional practitioner. These include the traditional midwives and some herbalists.[16]

The integration of such a heterogeneous group of fully trained, half-trained, and untrained practitioners into any official health care system poses many problems. It should be borne in mind, however, that traditional practitioners far outnumber modern health professionals and that over 90 per cent of them work in the rural areas, whilst the remainder cater mainly for the disadvantaged populations in urban areas. With regard to integration, the first group with fully integrated training pose no problem. They are perhaps the ideal practitioners in a traditional society. They seek out and utilize the best practices in both systems, and the well-established and experienced ones often have lucrative consultant practices in big cities or are engaged in research activities. The second category also pose no real problems. They practise almost exclusively in rural areas and some have been engaged in Government clinics in India, Sri Lanka, Burma, Nepal, and China with satisfactory results. Integrated practices have also been introduced into some central hospitals, where these traditional practitioners work alongside modern physicians and use traditional remedies, including herbal medicines. Their main function is in the curative aspects of health care, but preventive and community health practices can be inculcated through in-service training.

The third and fourth categories with no institutional or formalized training, however, pose several problems. The correspondence courses followed by the third group offer no practical training essential for the acquisition of skills. These are generally part-time practitioners—civil servants, teachers, agricultural officers, etc. Some preliminary orientation courses and in-service training would be required before any form of integration could be considered.

The traditional midwives constitute a special category. In the developing countries, they form the main body of primary health care workers in maternal and child care, and in some countries they are responsible for over 90 per cent of the births. Their role and functions have already been described in this chapter.

The integration of traditional midwives into health teams has already been started by several countries in Africa and Asia during the past two decades. Training programmes have been developed to improve their efficiency and to include family planning activities. Information on TBAs should first be obtained before planning any training programmes, with a view to integration. Data should include their role and functions, numbers and distribution within the community, and training including apprenticeship to established traditional midwives, approximate age and years of practice, etc. In the analysis of such data, the TBAs that would benefit from in-service training could be identified, and their effect on the health services also determined.

Orientation and training for TBAs require special consideration from the health authorities. TBAs, though illiterate and never exposed to any formal educational processes, are experienced, intelligent, and respected persons in the community. Their training should therefore be undertaken by health professionals, trainers, and supervisors of comparable or preferably greater age and experience, otherwise personality clashes and conflict would occur. The training should be mainly on the job and based on reinforcing useful traditional practices whilst simultaneously introducing new methods by demonstration but adapted to traditional ones. When special care or advice on more difficult problems are required, the primary health care worker should have access to more highly trained health personnel with adequate facilities. The types of such personnel will vary according to the resources of each country but they should have full appreciation of the various problems at the primary health care level to handle problems referred to them effectively. Their social and educational responsibilities in health education should be emphasized and their main educational function should be to teach the people how to look after themselves, with special attention to the mothers —'Educate a mother and you educate a family'. Health workers could first learn what mothers believe and feel and then help them to understand how they could improve their own and their family's health, mainly through practical demonstrations.

One major difficulty is to get townsfolk with their culture and education to understand country folk with their non-scholarly systems of belief. The effort of understanding must first come from the health authorities and the teachers.

Some serious thought will have to be given to referral arrangements, since without such back-up facilities the community would gradually lose confidence in their primary health care system. Referrals should be carefully explained to both the patient and relatives because in several traditional societies referral to hospital is taken as an indication of hopelessness and that the patient is being sent to hospital to die. Patients do not cherish being treated far away from their homes and so far as possible health interventions should therefore take place at community level.

A clear explanation of these interventions and the technologies used should always be given in simple language and preferably in the vernacular. Newspapers, films, radio, and television could all be used for mobilizing community interest and support for the development of the primary health care programmes.

In communities where health committees are established for the implementation of health programmes it should be ensured that not only Government officers and educated persons serve on these committees but that community leaders, literate and illiterate, and senior traditional practitioners and midwives are included as active participants.

Their opinion should be sought on the initial planning, health priorities, and overall implementation of the programme. It is mainly through such collaborative efforts entailing some devolution of responsibilities coupled with delegation of authority that enthusiasm for community participation can be assured. Traditional practitioners can be very useful allies in health work; on the other hand they can create a lot of difficulties at primary health care level—'If you cannot beat them, join them!'

Medicinal plants and herbal remedies

The practitioners of traditional and indigenous medicine rely mainly on medicinal plants and herbs for the preparation of therapeutic substances;

and the plant kingdom remains a treasure-house of potential drugs. The story of herbal medicines is a fascinating one: quinine, until recently the only effective remedy for malaria; morphia, the pain killer; digitalis for the failing heart; emetine for amoebic dysentery; ergot in midwifery practice; rauwolfia the tranquillizer, lunacy antidote, and antihypertensive agent; all of these were well known to the healers and medicine-men of Africa and Asia many centuries before their introduction into modern and scientific medicine, as were curare and penicillin, and are but a few of the natural products that have brought much relief to ailing mankind. In oncology, podophyllotoxins, the vinca alkaloids, colchicines, and some antibiotics are also derived from plants.

The high cost of drugs and the inability of many developing countries to purchase such drugs have prompted several countries to look for local products in the form of medicinal plants and herbal medicines that have proved to be effective, safe, inexpensive, and culturally acceptable. There are many records of traditional therapies using herbal medicines that are said to be very effective against common ailments and usually without any side effects. A preliminary task would be to identify the most commonly used formulations for the treatment of common ailments in various localities.

The identification of locally available and commonly used medicinal plants and herbs would have to be effected and a list compiled. In certain Asian and Latin American countries the cultivation of commonly used medicinal plants is undertaken in home and community gardens to ensure adequate and continuous supplies. The cultivation of medicinal plants and herbs could be conveniently linked with the production of vegetables and fruits with high nutritive value that should be of particular benefit to mothers and children. Such cultivation of commonly employed medicinal plants could also reduce the risk of extinction of the endangered plant species.

In the selection of drugs and herbal medicines for treatment, the health worker is expected to give preference to locally produced and equally effective preparations, and in the more established indigenous systems traditional practitioners certainly do this whenever possible.

A major constraint in the use of vegetable drugs is the absence of national pharmacopoeias in many countries. A few countries such as India and China have developed such pharmacopoeias, containing recipes of plant and herbal medicines. Others have old manuscripts which describe plants and herbs, including their properties such as taste, odour, and changes during digestion, potency, and specific therapeutic actions. In most other Third-World countries the systematic utilization of these medicinal plants in the pharmaceutical industry and for the purpose of health care has not been optimized and remains limited mainly to the production of traditional drugs for local use.

Different drugs are now being combined in one preparation to reduce adverse side effects—e.g. combining streptomycin with a Chinese traditional drug, and also the combination of active principles of Chinese, Indian, Japanese, and Arabic medicines in one mixture so as to obtain an effective but simultaneously smaller dose.

Primary health care workers should have basic knowledge on medicinal plants, their cultivation, identification, collection, and preparation for therapeutic applications within the community in which they work. The use of medicinal plants and traditional medicine could make people become more self-reliant.

The orientation of health professionals

Before existing manpower categories could be mobilized for maximum utilization, it would be necessary not only to give training and orientation to traditional practitioners as already outlined, but to ensure that all relevant professional health personnel receive appropriate orientation in the broad principles and practice of their indigenous and traditional systems. The early establishment of a dialogue among practitioners of the different systems in order to eliminate prejudices and to help them to develop more acceptable attitudes should be initiated at the highest possible level and include the leaders of the various practitioners and senior Government officials.

Ideally, orientation should be achieved early in the undergraduate period through multidisciplinary and integrated field projects including medical, nursing and midwifery students, pharmacists, sociologists, healers, traditional practitioners, and midwives at community level. Seminars and workshops could be organized for established practitioners and administrators and where free discussion could take place on various local health problems. Such activities have been operational at local, national and international levels since the early 1970s with salutary results. Much has been learnt by both modern and traditional practitioners, culminating in considerable mutual respect.

Conclusion

The promotion and development of traditional medicine can foster dignity and self-confidence in any community through self-reliance, and also reduce a country's drug bill considerably. What is more respectable than to take care of oneself within one's own means? The integrity and dignity of a people stems from self-respect and self-reliance, and traditional medicine *inter alia* can help promote this situation in many ways. Through carefully structured and well-designed in-service training and orientation programmes, the gap between the healers and western-trained physicians could be bridged and effective communication established. Integration of traditional and modern health services may be a difficult final goal but collaboration between the various health workers could be begun with immediate effect. Through such collaborative efforts, the various useful practices cherished by the majority of Third-World populations could be refined and reinforced by the application of modern science and technology, and the harmful practices eliminated. The western-trained physicians would also have much to learn through contact with traditional practitioners.

References

1. Bannerman, R. H. 'WHO's programme'. *World Hlth.* November (1977).
2. Razzack, H. M. A. *Principles and practice of traditional systems of medicine in India.* Ministry of Health and Family Welfare, Government of India (1977).
3. Maclean, U. *Magical medicine; a Nigerian case study.* Penguin, London (1971).
4. Bannerman, R. H. 'Traditional Medicine in modern health care services. International relations'. *J. David Davies Memorial Inst. Intl. Studies* **VI**, May (1980).
5. International Federation of Gynaecology and Obstetrics and the International Confederation of Midwives. *Maternity care in the world.* 2nd edn. (1976).
6. *Traditional midwives and family planning population reports*, series J, No. 22 (1980).
7. De Loundes-Vederese, M. and Turnbull, L. *The traditional birth attendant in maternal and child health and family planning.* World Health Organization, Geneva, offset publication No. 18 (1975).
8. Danfa Project. *Final report.* Ghana (1979).
9. Cosminsky, S. 'Cross-actual perspective on midwifery'. In *Medical anthropology* (eds. F. Crollig and H. Haley). Mouton Press, The Hague (1976).
10. World Health Organization. *Traditional birth attendants.* WHO, Geneva, offset publication No. 44 (1979).
11. Report of a WHO Seminar on Traditional Practices affecting the Health of Women, Khartoum-Sudan (1979).
12. Farnsworth, N. R. *World Health Organization—special programme in human reproduction.* Text of a lecture delivered at a Symposium organized by IUPAC, Washington (1979).
13. WHO, Geneva. WHO Technical Report Series No. 622 (1978).
14. Jayasuriya, A. and Fernando, F. *Principles and practices of scientific acupuncture.* Lake House Printers and Publishers, Colombo Sri Lanka (1979).
15. Kao, F. F. and Kao, J. L. (eds. *Recent advances in acupuncture research.* Institute for Advanced Research in Asian Science and Medicine, New York (1979).
16. Director-General of The World Health Organization and the Executive director of the United Nations Children Fund.
17. Mahler, H. 'People'. *Sci. Am.* **243** (1980).
18. World Health Organization. *The promotion and development of traditional medicine.* WHO, Geneva, Technical Report Series No. 622 (1978).

4 Culture and primary child health care

D. B. JELLIFFE and E. F. PATRICE JELLIFFE

A community's culture embraces all aspects of life, including religion, philosophy, and art. Whenever possible, it is most valuable to have as wide an understanding of any culture as possible. However, for practical reasons often only more limited areas of special interest of direct concern to child health can be looked into. Some of these, which have direct relevance to health or illness in young children and their mothers, are given in Table 4.1.[1-4]

TABLE 4.1 *Culture pattern and maternal and child health*

Pattern	
Parental	Child-rearing practices 'unusual' children (e.g. twins, albinos, etc.)
Maternal	All women, pregnancy, labour, puerperium, lactation *(doula)*
Young children	Newborn: cord, feeding (cf. colostrum) Transitional: diet, method of (weaning) separation
General	Disease classification: causation and prevention, treatment, healers Food ideology Miscellaneous: naming, significance of exact age, meaning of colours, etc.

All of these can be considered in relation to primary child health care, but the present paper will concern itself principally with one of the more important aspects of culture—'food ideology'.[4,5] Common categories include food/non-food; cultural superfood; 'body physiology' foods; age-group foods; 'sympathetic magic' foods; and prestige/celebration foods.

Cultural interaction

All contacts between health personnel and patients or the public are, in fact, cultural interactions. This is so, for example, when a paediatrician is dealing with a child and his parents in the physician's own culture. Likewise, when the public health practitioner is undertaking programmes in the community, cultural interaction occurs between what may be termed the concepts of westernized technical cultural and the *mores*, practices and attitudes of the community. Such cultural interaction is obviously most likely to be extreme when the western technical culture and the traditional culture are widely divergent.

Need for knowledge

A knowledge of the culture of a community is essential for health practice and can be useful in various ways (Table 4.2).[4]

TABLE 4.2 *Culture pattern: need for knowledge by health professional*

Pattern of disease (cf. rickets, tetanus neonatorum)

Preventive programmes:
acceptance of health education; identification of 'at risk' factors

Establishment of rapport

Increased world knowledge (cf. *doula* system)

Components of technical culture

Often the approach to cultural interaction by technical health workers is subconsciously viewed as a one-way flow—that is, with attention given only to the traditional culture and its components. It is underappreciated that the western technical culture has inherent ingredients and in the interaction these will need to be considered and play an important part.[6] Some of the main components of western technical culture, many of which are not often appreciated, are indicated in Table 4.3.

TABLE 4.3 *Aspects of 'Western' technical culture*

Numerophilia
Time-dominated
Curative allopathy (technical intervention)
Culturocentricity

Avoid; culturally inappropriate or

Generally harmful (cf. separation of mother/newborn)

Commercial cultural manipulation
Physician–hospital dominated
Technical preferred to biological

Incorporate; culturally appropriate
Generally benefical (cf. immunization)

Culture and primary health care

In the hospital, the interaction between technical and traditional cultural concepts takes place in an atmosphere which is predominantly technical. However, as health activities move peripherally, and especially in the village, so the interaction occurs with technical culture being introduced into a dominantly traditional cultural environment (Table 4.4).

TABLE 4.4   *Varying cultural interaction at different levels of health service*

	Technical culture		Tradition
Hospital	+ + +		+
Health centre	+ +		+
Village activities*	+		+ + +

* Primary health care.

It follows, therefore, that the village-level worker will be automatically working in a traditional cultural environment into which carefully selected technical cultural practices are being introduced. This poses certain inherent problems. First, the primary health care worker is of the village and may, indeed, be a believer in the traditional practices and may even need to be really convinced of the significance of the modern western concepts that he or she will be trying to introduce. Second, as ideas on cultural relativity are rarely taught in medical schools or other schools for health workers, the trainers of village health workers may have been very ill-prepared in this regard.

Practical approaches

A three-stage practical approach is often helpful (Table 4.5). It is necessary to investigate the local culture pattern by various means (always a continuing process) and to analyse and to categorize it into beneficial, neutral, uncertain, and harmful practices; appropriate action for the four main categories may then be suggested (Table 4.6).[4,7,8]

TABLE 4.5   *Cultural factors: a practical approach*

Investigate: literature, discussion, observation (task analysis)
Analysis: according to health consequences

Action		
	Beneficial:	Incorporate positively and prominently
	Neutral:	Ignore or incorporate unobtrusively
	Uncertain:	Observe → categorize later
	Harmful:	Dissuade (conviction) or integrate

TABLE 4.6 *Approaches to categories of cultural practice*

	Practice				
	Beneficial	Neutral	Uncertain	Integration	Harmful
Approach:	Incorporate (positively, prominently)	Ignore or unobtrusively incorporate	Observe and categorize later		Dissuasion (conviction)
Examples	*Doula* (female assistant at childbirth: India & most cultures); rest & good diet in puerperium (Chinese in Malaysia); prolonged breast feeding (most traditional cultures); weaning food vegetable oil with rice (Burma)	Avoidance of 'twin' bananas in pregnancy (E. Africa); symbolism of colour (yellow = danger/ill health), used in weight charts (Indonesia)	Cosmetic outlining of eyes; *Kajal* (carbon & oil), harmless; lead compound, harmful. Maternal pre-chewing of foods for infant; beneficial	Restriction of items for sick young child (hot/cold incompatibility)—use alternative food with hot/cold compatibility	Nutrition rehabilitation concept (mother's preparing foods and feeding) —observation important

However, often underappreciated is the need to undertake a similar mirror-image type of investigation into western technical practices in general and in particular in relation to the local ecological reality.

Cultural synthesis

Understanding of this interaction between western and traditional cultures includes analysis of technical culture itself and an understanding and awareness of its harmful and beneficial practices. Usually, the main features of cultural interaction in primary child health care in the village circumstances will be concerned with the introduction and acceptance of carefully selected beneficial technical practices, the avoidance and non-inclusion of harmful technical practices, the modification of harmful traditional practices, and the support of beneficial practices in the community. A scheme for this type of cultural interaction and synthesis is given in Table 4.7.

TABLE 4.7   *Cultural synthesis in primary health care*

Introduction of technical culture	Modification of traditional culture
Harmful practices *Avoid* introduction (e.g. separation of mother and newborn, promotion of bottle feeding)	*Modify* Cultural integration (substitution) Dissuasion by convincing demonstration
Beneficial practices *Introduce* appropriately modified: 　Cultural integration 　Persuade by convincing 　demonstration (cf. oral 　rehydration)	*Incorporate* Positively and prominently

Conclusion

The need for an awareness that medical and public health activities are cultural interactions requires widely emphasizing and should be part of the training of health professionals. It is particularly important that the training and activities of the primary child health worker in the village be based on an awareness of this inevitable interaction and the need for cultrual synthesis.

References

1. Jelliffe, D. B. and Bennett, F. J. 'Indigenous medical systems and child health'. *J. Pediatr.* **57**, 24 (1960).
2. Jelliffe, D. B. 'Culture, social change and infant feeding'. *Am. J. clin. Nutr.* **10**, 19 (1962).
3. Jelliffe, D. B. and Bennett, F. J. 'Cultural and anthropological factors in infant and maternal nutrition'. *Fedn. Proc.* **7**, 185 (1961).

4. Jelliffe, D. B. *Infant nutrition in the subtropics and tropics*, 2nd edn. WHO, Geneva, Monograph No. 29 (1968).
5. Jelliffe, D. B. 'Parallel classifications of foods in developing areas and industrialized regions'. *Am. J. clin. Nutr.* **20**, 279 (1968).
6. Jelliffe, D. B. and Jelliffe, E. F. P. 'The cultural cul-de-sac of western medicine (towards a curvilinear compromise?)'. *Trans. R. Soc. trop. Med. Hyg.* **71**, 331 (1977).
7. Jelliffe, D. B. and Jelliffe, E. F. P. *Human milk in the modern world*. Oxford University Press, Oxford (1978).
8. Jelliffe, D. B. 'Cultural blocks and protein malnutrition in early childhood in rural West Bengal'. *Pediatrics* **20**, 128 (1957).

5 Focused national nutrition surveys

A. J. ZERFAS, B. BROWDY, W. D. CLAY, I. J. SHORR, D.B. JELLIFFE, and E. F. PATRICE JELLIFFE

Nutrition surveys in developing countries have been criticized as irrelevant, inefficient, and inconclusive leading to little subsequent change in nutrition. Latham, for example, remarked at a recent seminar on nutrition planning: 'Nutrition in Africa might be much better if children could digest cellulose and could eat all the survey reports and non-implemented nutrition plans that have been produced.'[1]

However, nutrition surveys can be of value if careful attention is paid to objectives and outcomes.

In the past 30 years, two major types of surveys have been undertaken. Although some were longitudinal,[2-5] most were cross-sectional[6-13] because of reduced cost, time, and complexity. Even the cross-sectional type varied greatly in content, expense, coverage, and attention to sampling.

Minimally, a nutrition cross-sectional survey should supply the point-prevalence of malnutrition (that is, the number and percentage of malnourished individuals at that particular time) in the population to be described, emphasising at-risk groups, mainly young children. However, such an investigation provides only a 'snapshot' picture of nutritional status at one season in a particular year.[7,14]

Nutrition surveys providing national data can offer unique administrative advantages for policy making. Obvious disadvantages are the relatively high cost and effort required, even over a comparatively short period. When a national nutrition survey is under consideration, certain questions demand realistic and, at times, humble replies:

1 Who requested the survey and why?
2 Is this survey really necessary?
3 Are there cheaper, more efficient methods of obtaining the information?
4 Can teams complete the survey within the known and likely constraints, including those of finance, personnel, geography, and climate?
5 Will existing or future national nutrition policies incorporate the reviewed results and recommendations?
6 Will the Government use people trained for the survey for future studies, such as surveillance or continuing screening programmes for health care?

In this account, focused national nutrition surveys (FNNS), refer to those cross-sectional surveys with a relatively simple, low-cost methodology based primarily on anthropometry and haemoglobin estimations in young children, with careful attention to sample design and sample size to ensure an adequately representative description of nutritional status throughout the nation. Several of these surveys have recently been completed.[15]

This chapter outlines the background and methods of FNNS, based on experiences in developing countries. The results are presented as examples of the types of information derived from such surveys. It would be premature to discuss fully the policies arising from these results. Moreover, some of these policies require knowledge outside the scope of FNNS, such as the macro-factors responsible for influencing levels of undernutrition throughout the country.

Core items

Certain key measures have made up the 'basic core' of such surveys (FNNS): weight, height (stature), precise age, and oedema for detecting protein-energy malnutrition (PEM) in pre-school age children; haemoglobin estimations for anaemia in these children; and qualitative dietary recall for feeding patterns. If prior information suggested that specific deficiencies were of national concern, clinical signs, such as goitre in mothers for iodine deficiency, were included (Table 5.1).[16,17]

TABLE 5.1 *Survey items**

Category	Items	Method/equipment
	Core items	
Identification	Child number	
	Child name	
	Person interviewed	
	Sex	
	Age	Birth record
		Local calendar of events
Anthropometry	Weight	Salter hanging scale
	Height–length	Height–length board
Clinical signs	Oedema	
Laboratory	Haemoglobin	Cyanohaemoglobin method
Dietary	Qualitative 24-hour recall (including breast feeding)	
	Non-core items	
Demographic	Household size	
	Who takes care of child	
	Birth order	
	Birth interval	
	No. of wives	
	Mother pregnant, lactating	
	Tribe mother/father	
	Religion mother/father	

TABLE 5.1—*continued*

Category	Items	Method/equipment
	Non-core items	
Anthropometry	Arm circumference	Zerfas insertion tape
	Triceps fat-fold	Tanner/Whitehouse caliper
	Maternal height, arm circumference, fat-fold	As for child
Laboratory	Haemoglobin of pregnant mother	Cyanohaemoglobin
	Measles antibody	Filter paper method
	Malaria	Thick and thin film
	Blood picture	Thin film
	Sickling	Hb electrophoresis
	Stool parasites	Qualitative film
Morbidity	Clinic attendance	Clinic record
	Recent illness: diarrhoea, fever	
	Temperature	Rectal thermometer
	Dehydration	Loss of skin turgor
	Measles history	
Mortality	N. died of measles	
	Total deaths	
Dietary	Breast feeding	
	Bottle feeding	
	Sources of milk	
	Weaning foods	
Socioeconomic	Occupation father/mother	
	Mother able to read	
	Type of house	
	Possessions (e.g. radio)	
	Method of cooking	
	No. and type of animals	
	Source of water	

* All items pertain to child unless otherwise indicated. When required, include clinical signs for vitamin A or iodine deficiency, pellagra, and rickets.

Non-core items

'Non-core' items—such as indicators of socioeconomic status, morbidity, and mortality—were selected to provide some indication of the background of the population, in terms of possible relationships with health and nutritional status. Although desirable, such items are often subjective, difficult to define, and prone to recall errors. Moreover, their relationships with undernutrition could not be interpreted as causal, because of the cross-sectional nature of the survey. Hence, such items had a lower priority for inclusion in the surveys. As the term FNNS indicates, such surveys have 'focused' mainly on prevalence rates of PEM and anaemia obtained by locally trained field workers, in a randomly selected national sample.

Origins

After the 1974 World Food Conference, the United States Agency for International Development (US AID) through its missions advised certain Governments about available US technical assistance for cross-sectional nutrition surveys, with special reference to young children. The results were intended to provide baseline data on nutritional prevalence as a necessary but incomplete part of nutrition planning, including the development of nutritional surveillance.

Contractors and countries. US AID contracted two institutions, the Center for Disease Control (CDC), Atlanta, Georgia and the University of California, Los Angeles (UCLA), to respond to Government requests. From 1975 to 1978, nine surveys based on the FNNS methodology were undertaken. CDC collaborated with the Governments of Nepal, Sri Lanka, Togo, Haiti, and the United Arab Republic;[18-24] UCLA with the Governments of Liberia, Lesotho, Sierra Leone, and Cameroon (also called the Cameroons).[25-28]

Organization and schedules. Countries indicated their interest for a survey to the local AID Mission, who notified the Office of Nutrition, US AID/Washington, who in turn informed CDC or UCLA. Subsequently, the US contractor visited the country which requested a possible survey. During this site visit, if Government officials, AID Mission, and contractor decided favourably on a forthcoming survey based on the FNNS method, responsibilities, scope of work, funding, and schedules were defined.

After an interim planning period of 3–6 months, the US team returned for the training and field phase, which lasted from 3 to 4 months. The contractor then analysed the data and wrote the report in the USA, assisted at times by visiting professionals from the country which had been surveyed. About 6 months after the completion of the field phase, a draft report was presented to the respective country Government, by the combined national and foreign teams. The final report with recommendations was prepared after this review process.

Differences between CDC and UCLA approach. Both CDC and UCLA used the same basic methodology, yet there were certain differences. Surveys in countries with UCLA involvement usually contained more items outside the 'basic core' (see above). These varied, but included weaning patterns, arm circumference, and triceps fat-fold measures in children, anthropometry in mothers, retrospective mortality data, and information on the prevalence of certain parasites. Government requests influenced item choice. Surveys in countries with CDC involvement usually had less survey items, but a larger total sample size.

African surveys with UCLA collaboration

This paper briefly describes the national surveys in four African countries—Cameroon, Lesotho, Liberia, and Sierra Leone—in which UCLA assisted (Fig. 5.1 and Table 5.2) in order to draw some general conclusions from them for wider use. For the most part, these countries were sparsely populated and had road transport difficulties. Many sites were accessible only by foot, horseback, boat, or air. Two countries, Sierra Leone and Lesotho, had a prior national nutrition survey, each done some 20 years previously.[29,30] In Cameroon, previous dietary and anthropometric studies had shown regional differences in the nutritional status of young children and adults.[31]

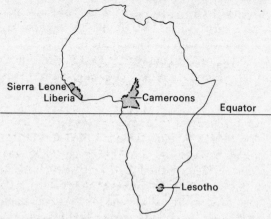

FIG. 5.1. African countries surveyed with UCLA collaboration.

TABLE 5.2   *Countries surveyed*

Country	Population (million)	Density (per km²)	Major tribes	Capital city	Population (thousand)
Liberia	1.5	13	Kpelle Bassa	Monrovia	120
Lesotho	1.1	36	Basotho	Maseru	25
Sierra Leone	3.3	118	Temne Mende	Freetown	200
Cameroon	7.7	16	Bamilike Foulbe Beti	Yaounde	350

Sources: Area handbooks for Liberia (1972) and Cameroon (1974); Foreign area studies, Government Printing Office, Washington D.C., 20402; Africa Guide (Ed. R. Synge), Africa Guide Company, Essex, England (1978); Provisional census data for Liberia (1974), Lesotho (1976), Sierra Leone (1974), and Cameroon (1976).

Senior direction

Both the country senior Government professional and the UCLA representative shared technical direction for the field survey, while a national

Government official was administrative director. In two countries, the Ministry of Health was the responsible Government agency; in the other two, the Ministry of Planning.

Funding

Governments of the countries surveyed provided funds and support for transport, salaries of local field personnel, and services. US AID/Washington supplied other in-country funds and all UCLA survey costs. Contractual amendments for UCLA were required yearly and after each site visit, the latter to clarify the projected costs for the individual countries. Although necessary, the time taken for these new contractual agreements often compromised schedules, particularly for the field phase of the survey.

Costs

It is difficult to estimate costs for the individual surveys since the planning, field phase, and analysis overlapped for the four countries. Furthermore, unexpected delays resulted in added project costs unrelated to the surveys. For example, the Sierra Leone Government (for reasons unrelated to the survey) required the scheduled field phase to be postponed. As a result, no surveys were done for several months, since there was no alternative country requiring a survey at the time. Consequently, total costs may be over-estimated by as much as 20 per cent.

The average overall cost for each survey was about US $180 000 (range US $120 000–280 000). This includes travel between the USA and Africa, salaries of senior professionals, and all support costs at UCLA. Of this, Governments of countries surveyed contributed some 20 per cent. Total field costs averaged US $90 000 (range US $45 000–180 000). The estimated expense for each sampled child and mother was US $15. All expenditures are based on estimates and are not quoted from any publication.

Behar[32] reported an average cost of US $160 000 for a cross-sectional survey in a Central American country during 1968–70. Schaeffer[33] stated that the 35 ICNND (Interdepartmental Committee for National Nutrition Defense) surveys conducted in developing countries from 1955 to 1967 cost US $8 million, or an average of US $230 000 each.

Survey design

Population selection

For all four countries, a recent (within 1–3 years) national census provided the population data from which the sample was selected (sampling frame). This sampling frame included the entire country, apart from Liberia; where 5 per cent of the rural population could not be included in the survey owing to inaccessibility (no available transportation of any kind within the time constraints). In that same country, only shanty areas of the capital, Monrovia, were selected, since these were the major areas of concern in an urban setting, where adequate sampling was difficult.

Countries were stratified (i.e. partitioned) by major political divisions, such as provinces; and by capital city, urban areas other than the capital, and rural areas. Where relevant and feasible, the survey included further descriptions, such as those defined by the World Bank for Cameroon—sahelian (north), central savannah, west plateau, west plains, and south rain forest.

Sample design

Sampling followed a stratified multistage cluster design. For the first stage, a random start systematic sample of enumeration areas (EAs) was used. An EA is a geographically defined population, generally between 500 and 1500. Within the EA, one site (defined as one or more villages, depending on size) was selected with probability proportional to population. Within each site thirty children were selected. In Sierra Leone, this first stage was coordinated with an ongoing national household expenditure survey[34] and in Liberia with a current agricultural survey (in rural areas).[35] This was done to link results, facilitate sampling, and reduce costs.

The sampling of children within each site varied among countries. The choice of method depended on several factors, including available maps, reliability of lists of heads of households, anticipated migration patterns since the last census, and above all the ability of the surveyors to carry out instructions adequately with limited or intermittent supervision. Where possible, teams used household lists for each site. If such lists were unavailable, other methods, such as the numbering of household structures, were done. In all cases teams strived for random selection of households.

In Sierra Leone and Lesotho, senior personnel, based on advice from local and US statistical consultants, usually prepared household lists or maps (from census, or other surveys or sources) before teams visited the sites. They also identified the first household, using a random table of numbers and the pre-determined direction teams would take (following the list of names or map).

On arrival at a site a team and supervisor, when available, checked the list and map with the village chief and assistants. Then teams visited the first household selected. If this household had an available child aged 3–59 months, the child was measured. Teams continued to survey households as directed until they measured thirty children. This total provided about six children per site for each year of age.

In Cameroon and Liberia, after teams arrived at a site, they prepared maps which delineated dwellings (or groups of dwellings). In these countries, the first household was randomly selected on a basis of geographic location and subsequent ones visited in a predetermined direction.

Sample size

Thirty sites for each area to be described, making a total of 900 children, were considered sufficient for a precise estimate of the prevalence of malnutrition.[36,37] For example, with this sample size, there is an 80 per cent chance of finding a significant difference (where $\alpha = 0.05$) in undernutrition prevalence rates if in fact they are 20 per cent and 13 per cent in the two areas.[38]

TABLE 5.3 *Survey sampling problems*

Types	Examples	Estimated frequency (%)
I Non-coverage	Failure to identify sample sites (e.g. due to inaccessibility)	0.7
II Non-response	Temporarily absent (market, festivals, visiting relatives, in fields)	Unknown (? 5–15)
	Refuses to participate	Rare
	Too ill to measure	Rare
	Measure unreliable (e.g. age)	1–5
III Faulty data	Exclusion of data from analysis Incorrect age of child Absent or impossible weight or height measures	1
IV Inadequate sampling frame	Population change since census Migration (rural to urban, seasonal) Births, deaths	Unknown

Sampling errors

Sampling errors had varied origins and were often difficult to measure (Table 5.3). Problems of geographical non-coverage were rare. Teams visited every site in all countries but one. The exception was Lesotho, where teams could not survey three of the twenty mountain sample sites, nor could they visit randomly preselected alternative sites. Impassable rivers, due to unusually heavy and early rains, forced teams to choose sites as near as possible to those originally selected. In this case, tight schedules prevented suitable senior supervision or a re-survey of the proper sites.

Examples of the team difficulties in reaching and surveying selected sites abound. In Liberia, a team waited 4 days until villagers returned from a funeral. In Lesotho, a team rode horses over mountainous terrain for 4 days to reach a site. A further problem in Lesotho was that village names may have changed when a new chief was appointed.

Children on farms distant from the village were less likely to be included in the sample, due to the method of sampling and to inevitable logistical problems. Some survey evidence suggested that remoteness was positively related to higher prevalence rates of undernutrition. Hence, results may be an underestimate of the true prevalence of malnutrition, especially in more rural areas, while differences between urban and rural areas may also be underestimated.

Problems with incomplete enumeration

Teams could not enumerate all children aged 3–59 months in the site, because the lengthy process of age determination would compromise the relatively short time (2–3 days) allocated for each site. Thus, an accurate estimate of non-response could not be determined.[38]

Weighting (statistical adjustment) should compensate for known, but different, probabilities of child selection for each site, but incomplete enumeration prevented such weighting. As the first sample stage was based on total population and the second stage on children aged 3–59 months, such probabilities are derived from the ratio of the number of these children (including non-respondents) to the total population per site.

Instead, the analysis assumed that the ratio of young children to total population was fairly constant among sites (based on recent census data, this ratio was about 0.16 for a whole country). If this assumption of constant ratios was not true, then sites with a higher proportion of the sample children per total population would be over-represented in the results. How this potential bias affects the true undernutrition prevalence rates is unknown, although its influence is probably small.

Survey options

Sampling within sites demands more time for adequate enumeration and follow-up procedures, particularly for non-respondents. This means adding costs or reducing other survey elements, such as sample size or number and complexity of items for the same total cost.[39]

Survey members moved 'caravan style' throughout the country. Each of the 8–10 teams usually included two surveyors. The 'caravan method' ensured optimal communication in the field among teams, supervisors, senior survey personnel, and vehicles. Also, unknown errors by the same team would tend to be repeated from one region to the next. This would minimize the effect of these errors on results comparing regions.

Although this caravan approach has advantages, it may limit the total period for planning, survey, and follow-up within a particular region of the country. In contrast to the caravan method, the World Fertility Survey in the Cameroons assigned each team to a specific area in the country for the field duration. This survey had 1 year for intensive in-country preparation, had teams comprised of five surveyors, and included a questionnaire with numerous attitudinal items, with prolonged field work.

Training

During the first 1–2 weeks of the field phase, senior professionals finalized preparations for the survey design, the choice of items, the logistics, and the training, including completion of the training manual. Thus local professionals were sufficiently briefed to play a major part in the training proper.

Trainees were usually regular employees of the Ministry of Health, primarily health assistants working in various regions of the country. Assignment was based on availability, past field experience, and language capabilities. The number of trainees were usually well above survey requirements, to allow for drop-outs and widen the choice for selected surveyors.

Training took place in the capital city and lasted 3 weeks. The training manual, prepared by the contractor after consultation with the representatives of the surveyed country, emphasized anthropometry, methods of age determination, interview techniques, the filling in of questionnaire forms, and survey protocol. Most of the training was devoted to practice and review. Sufficient time was allocated for trainees to attend to personal matters before the survey proper.

Surveyors were selected primarily because of motivation and performance. Supervisors were chosen either before the training on a seniority basis or after training, according to merit.

Teams conducted limited field trials in conveniently located villages in both urban and rural areas. From these, final modifications in the questionnaire and field procedures were made.

Anthropometric reference data

Anthropometry (such as weight and height)—being objective, practical, and relatively sensitive—is the preferred method for assessing prevalence rates for PEM in young children. Reference levels are needed to interpret such measurements by means of appropriate indices.

Selection of 'international' reference data

Data analysis was undertaken using the NCHS/CDC anthropometric reference levels to convert weight and height measures to indices, such as weight-for-height, height-for-age, and weight-for-age. These levels, recently recommended by WHO,[41,41] were based on two US sources: on the Fels Research Institute Study in Ohio (for children from birth to 3 years), and on the recent National Center for Health Statistics (NCHS) US surveys (for those from 2 to 5 years).[42] The Center for Disease Control (CDC) prepared reference data for arm circumference-for-height and triceps fat-fold-for-age, derived from the US Ten State Survey[43] and that of arm circumference-for-age from London children.[44] Such reference data plainly can provide only a basis for comparison, but should not imply a standard of normality or excellence.[45,46]

Special group studies

In all but one of the African countries surveyed, teams measured a special group of 'élite' children (up to 100 children for each year of age) with parents of relatively high socioeconomic status. Such children apparently had a better diet and health care than those throughout the nation, and thus their nutritional status was arbitrarily presumed to be 'optimal'. The exception was Liberia, where 'élite' children in Monrovia had greatly different ethnic and cultural backgrounds compared with those in the country in general (Table 5.4).

TABLE 5.4   *Special groups*

Country	Location	Source	Major age range (months)	Total number
Liberia	Rural	Mining area	6–47	212
Lesotho	Capital	Housing area	3–59	293
Sierra Leone	Capital	Nurseries	24–59	361
Cameroon	Major city	Housing area	3–59	505

In Sierra Leone, various crêches and nurseries in Freetown, the capital city, provided most of this special group. In Cameroon, teams measured children in selected housing estates of Douala, the major commercial city, where business or Government officials resided. In Lesotho, the approach was similar to that used in Cameroon. However, the relatively small population of Maseru, the Lesotho capital, limited the choice of subjects. In Liberia, the special group comprised children of regular workers living in a large mining-company housing settlement. These rural children had had ready access to health care since birth.

In three of the four countries surveyed, there were several major tribal groups, many of which differed genetically and culturally from those in the special group. The exception was Lesotho, where almost the entire population is Basotho, of Bantu origin.

Results for the special groups were compared, both with the US anthropometric reference levels referred to earlier (p. 65), and with the national sample. In the special groups for Sierra Leone and the Cameroons, the prevalence rates for all undernutrition indices except wasting closely approximated the USA reference population (Table 5.5). In the surveys of Lesotho and Liberia, these prevalence rates for the special group were generally two to three times greater than the US reference population but still far less than the national sample.

Relevance of reference data

The statistical median (50th centile) for the reference data is often loosely

TABLE 5.5 *Prevalences (%) of undernutrition in special groups*

Country	Type of undernutrition		
	Wasting*	Stunting†	Arm‡ Wasting
Liberia	2.9	9.0	4.2
Lesotho	0.0	11.3	5.8
Sierra Leone	0.3	2.1	0.0
Cameroon	0.0	4.2	0.8
Reference population ‖	1–3	1–3	1–3

* Wasting: under 80 per cent of reference median weight-for-height.
† Stunting: under 90 per cent of reference median height-for-age.
‡ Arm wasting: under 82.5 per cent of reference median arm circumference-for-age.
‖ Expected results from the US reference population.

termed the 'reference level' (or in previous literature the 'reference standard') or 100 per cent. Levels below or above are then designated as percentages (e.g. 80, 60, or 120 per cent). The growth of pre-school children in developing countries is considered to be associated far more with social and environmental factors than with those based on genetic differences.[47] Recent findings from the International Biological Programme suggest that genotype differences in growth—at least as expressed by weight and height of young children—between most Africans and most Europeans are slight.[48]

Certain problems in the interpretation of 'international' reference data when applied to developing countries include:

1 Although genetic differences are usually less than environmental, they do exist and should be considered. Moreover, the implications of reference levels vary with socioeconomic status.[49]
2 International reference data may have more than one origin, as in the present case, where the arm circumference levels were from London children and weight levels from children in the USA.
3 Investigators are divided among those preferring to continue with the 'old' Boston reference data because of ease of comparability with past results, those now using the 'new' NCHS/CDC data (endorsed by WHO) and yet others favouring one of several other sources.[45,49]
4 Furthermore, reference data may be interpreted in several ways. Recently, Waterlow *et al.*[40] recommended that results should be presented in the form of standard deviations and centiles below the median, rather than in the usual form based on the percentage of median reference values.

5 The designation of cut-off points indicating levels suggesting malnutrition may vary.[50]

The choice and method of analysis of reference data depends on a judicious mixture of convenience, comprehension by the non-mathematical, statistical exactitude, and, above all, practical utility.

Despite the problems and confusion, reference levels do relate to measures of morbidity and mortality. For example, follow-up studies in India[51] and Bangladesh[52] have shown that young children with anthropometric levels below certain cut-off points have a much greater risk of dying than do those above them.

Anthropometric indices

Weight-for-height ('acute undernutrition')

Low weight-for-height indicates 'wasting' or 'acute undernutrition'.[53,54] For each country, the prevalence of low weight-for-height (under 80 per cent of the reference median value) varied from 1.0 to 3.1 per cent, a relatively small number of affected children. During a famine, wasting may occur in at least 10 per cent of young children.[55] The prevalence of oedema, a sign of severe acute undernutrition (kwashiorkor), was also low, ranging from 0.1 to 0.6 per cent.

A single cross-sectional survey has certain disadvantages in interpreting the occurrence of acute undernutrition. These include:

1 It will be less likely to detect conditions of short duration.
2 To attain the required precision, the sample size for these surveys for uncommon events (e.g. under 2 per cent) is usually much greater than the recommended 900.
3 Incidence rates cannot be derived—that is, the number of new patients with malnutrition over a defined period (usually 1 year) divided by the population at risk.

The field surveys started when the rains ceased and sites became reasonably accessible. This was, therefore, not during the 'hungry' season, when acute undernutrition is probably at its peak.[56] Indeed, results for acute under-nutrition prevalence obtained during a 'nutritionally better' time of the year may mislead policy makers about the numbers at risk and may not identify priority groups affected during the hungry season.

Height-for-age ('chronic undernutrition')

Low height-for-age (retarded growth, nutritional dwarfing, or stunting) is considered indicative of 'chronic undernutrition'.[53,54] However, current stunting cannot identify the onset, duration, and recurrences of previous episodes of poor nutrition.[57] Prevalence of stunting rather gives a general indication of long-term 'unfavourable nutritional environment' during unspecified periods in the past. In this sense, stunting is an adaptation, at least as regards immediate metabolic breakdown ('clinical malnutrition') and hence a need for priority social or public health action.

Survey results paralleled the expected socioeconomic gradient within each country: élite, capital city, urbanized and rural from least to highest prevalence of stunting (under 90 per cent of the reference median value of height-for-age) (Fig. 5.2).

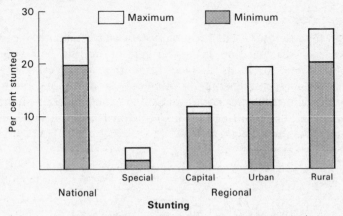

Stunting

(Under 90% of reference median height-for-age)

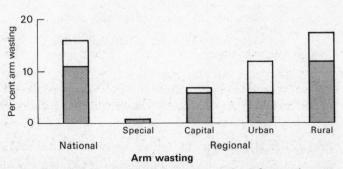

Arm wasting

(Under 82.5% of reference median arm circumference-for-age)

Fig. 5.2. National and regional comparisons for prevalence of stunting and arm wasting (Liberia, Sierra Leone, and Cameroons combined). Special group excludes Liberia.

Some results indicated that stunting was more strongly associated with environmental than with genetic factors. In Sierra Leone, stunting was found in 27 per cent of rural Mende children and in only 2 per cent of Mende children living in the capital city. This may be attributed to better diet, health services, and socioeconomic status in the city. Furthermore, the correlation coefficients between the heights-for-age for children and the heights of their mothers were low for each country (approximately 0.2), suggesting only a weak relationship between young child growth and genetic factors.

Recommendations to Governments based on the prevalence of stunting cannot give any guidance as to the relative importance of causal factors, although problems are assumed to be of long duration. Such problems may relate more to education and socio-economic factors than to diet and infections. Because stunting does not signal a need for urgent action, necessary interventions might be postponed.

Combined stunting and wasting

Waterlow has proposed a classification system which combined the presence or absence of stunting (low height-for-age) or wasting (low weight-for-height)[54] (Table 5.6). In its simplest form, this system categorizes four possible outcomes: stunting only, wasting only, both stunting and wasting, and normal (neither stunting nor wasting). For example, a child with stunting only has chronic undernutrition but no recent onset of acute undernutrition, as defined by anthropometry.

TABLE 5.6   *Waterlow classification*

	Not stunted	Stunted*
Not wasted	Normal	Stunted only
Wasted†	Wasted only	Stunted and wasted

* Stunted: under 90 per cent of reference median height-for-age.
† Wasted: under 80 per cent of reference median weight-for-height.

Because the prevalence rates for wasting were very low, this system contributed little of value to these particular survey analyses. Moreover, cross-classifications may confuse the interpretation of single anthropometric indices.

Weight-for-age ('underweight')

Low weight-for-age cannot indicate the duration of PEM, but does present a single *composite* prevalence rate which summarizes undernutrition status in young children. In addition, local nutritionists are more familiar with this measure, while complementary weight data from other sources, such as growth charts, are more readily available than data based on height.[58] The measurement of height (and length particularly) is often difficult to perform adequately and often requires specially designed boards. However, low weight-for-age includes two forms of malnutrition; acute and chronic undernutrition.

Arm circumference

The measurement of the arm circumference adds valuable information to descriptions of nutritional status. Arm-circumference-for-age is far less affected by age errors than are weight- or height-for-age, because its reference median values (reference levels) are almost constant from 1 to 5 years of age.[59] Hence, it may help to identify large errors in age and anthropometric measurement. Furthermore, especially with the insertion tape,[60] measures can be sufficiently precise.

As with stunting, there were marked regional differences in prevalences of 'arm wasting', defined as less than 82.5 per cent of the reference level—that is, the reference arm-circumference-for-age median value. This figure approximates to two standard deviations below the median (Fig. 5.2).

Arm circumference is particularly relevant when weight, height, and precise age measures are neither feasible (because of lack of funds or equipment) not interpretable (because of uncertainty concerning exact age).[61] This method can be used for screening, surveys, or surveillance with an appropriate 'categorizer', such as a Shakir–Morley strip or armband, which can even be made from strips of used X-ray film.[62]

Age determination

The precise age of the child (or birth date) is essential for age-dependent anthropometric indices, such as height-for-age and weight-for-age. Methods for age determination include verifiable birth record, clinic record, local calendar of events, and age stated by the mother or respondent.

In three of the four African countries surveyed, teams obtained documentary evidence of birth date in less than half the sampled children. This is a common problem in cross-sectional studies in developing countries. The development of a local calendar of events based on village information was emphasized in training and in use in the field.[63]

Informal field checks and data editing during preliminary analyses indicated that age errors were not uncommon. Such editing reviewed the agreement of the calculated age from the birth and interview date, grossly abnormal age-related anthropometric results, and transcription errors. Between half and one per cent of subjects had age errors serious enough to warrant exclusion from the analysis. Age errors of a less serious nature, but still distorting the results of anthropometric age-related indices, appeared far more often.

Age errors may misclassify a screened child above or below the under-nutrition cut-off point. In surveys, only a systematic bias—that is, a consistent over- or under-estimation of the true value—would result in a spurious increase of decrease in the prevalence of children reported with under-nutrition. However, indirect methods—such as comparing the age distributions for children with and without birth records—suggested that there was no serious systematic error in age determination (Fig. 5.3). Other African surveys, for example in Kenya, have shown similar findings.[64]

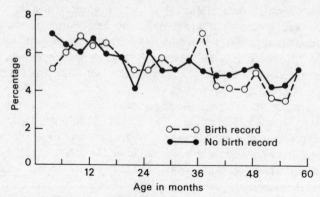

FIG. 5.3. Age distributions—Cameroon national nutrition survey, comparison of groups with and without birth record.

In young children, weight and height vary much more with age than do arm measures. Hence, age errors distort weight-for-age and height-for-age results more than they do arm circumference-for-age. A 3-month error for a child truly 18 months of age is equivalent to a 5 per cent error for weight-for-age, 3 per cent for height-for-age, and only 0.6 per cent for arm circumference-for-age.

Anaemia

Anaemia, especially in areas of high malarial endemicity, is only a non-specific indicator or 'suggestor' of possible dietary iron/folate deficiency.[65] Anaemia prevalence in children aged 6–59 months in three West African countries—Liberia, Sierra Leone, and Togo—ranged from 58 to 62 per cent. Lesotho, a malaria-free country due to high altitude, had a much lower prevalence (25 per cent). It is difficult, however, to assess the comparative importance of other factors causing anaemia, such as particular nutrients in the diet, genetic abnormalities (e.g. sickle cell disease), and parasites other than malaria, especially hookworm infestation.[66]

Maternal goitre

The prevalence of visible goitre (Grade II or III by WHO criteria)[67] of all mothers of sampled young children was 4.4 per cent in the Cameroons and 5.2 per cent in Lesotho. In the Cameroons, results added evidence and impetus to a previously postponed iodized salt fortification programme. In Lesotho, a WHO-assisted national survey (using similar diagnostic criteria) in 1960[30] found a goitre prevalence of 15–20 per cent. The reduction since then is probably related to iodized salt. As expected, there were large regional differences. Mapping of goitre prevalence by site clearly illustrated variations within and among regions.

Vitamin A deficiency

This was not investigated in any country, since pre-survey information suggested that prevalence rates would be too low to warrant inclusion in the surveys.[68]

Feeding patterns

Breast and bottle feeding

In all four countries, breast feeding was reported in 80–90 per cent of children aged 12 months. In the two countries investigated, the Cameroons and Sierra Leone, bottle feeding occurred in 8.8 and 20.7 per cent of all sampled children aged 3–5 months; in the capital cities the prevalences were 32 and 61 per cent, respectively.

Food other than milk

In all countries, 15–20 per cent of children aged 6–11 months and 5 per cent aged 12–17 months received no food other than milk (breast and non-breast milk). By the end of the second year of life children invariably ate the same foods as their mothers.

Weaning and family foods

Whereas the child may eat the same food as the family (family food) weaning food represents food especially prepared for the child during the weaning period. Fig. 5.4 illustrates these feeding patterns for North province Cameroon children according to age, but at one point in time. Breast feeding and weaning food (usually millet-based) consumption was still common during the second year of life, whereas that of family food was less common compared with other provinces. Because this province had the highest prevalence of underweight and arm wasting for the whole country the above findings require further investigation, perhaps in a longitudinal study.

Qualitative dietary recall

Qualitative dietary recall ascertained the types of foods eaten during the past 24 hours by children aged 3–24 months and their families. Recommendations of the National Academy of Sciences, USA[69] determined the basic methodology, which was then adapted to individual country requirements. This food recall method was qualitative, based on food patterns and not on quantities consumed.[70]

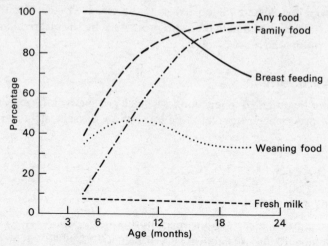

FIG. 5.4. Percentage of children consuming breast milk, fresh milk, weaning, family, and any foods in North Province, Cameroon. Weaning food is that especially prepared for the child; family food that shared with the family; any food, a combination of weaning or family food.

Food categories

Foods were considered separately or in 'categories'. The categories selected were those appropriate to the main foods of the diets in the four African countries surveyed and were abbreviated as follows: *tubers* (including roots such as cassava and starchy fruits, e.g. plantain), *cereals* (cereal grains), *vegetable proteins* (protein-rich plants—legumes and groundnuts), *animal proteins* (protein-rich animal products—fish, meat, and eggs), *milk* (breast milk, cows' milk), *DGLV* (dark green leafy vegetables), *fruits*, and *oils* (including fats, such as margarine).

Child–family comparisons

The comparative use of foods by young children and their families can be demonstrated in a bar diagram (Fig. 5.5). This shows the percentage of children (*C*) and their families (*F*) which consumed at least one member of a food category (such as oils) during 24 hours. Because 90 per cent of families consumed any oil, at least one item of that food category was potentially available to the child in these families. However, 40 per cent of children aged 6–11 months did not consume any oil, in the same families which ate that food.

Such 'deficits' may be considered as differences (as shown in Fig. 5.5) or proportions. Thus the percentage of families consuming a food category item provides a crude index of food availability, whereas the 'deficit' indicates an intra-family food distribution pattern—that is, foods available but under-used in feeding young children.

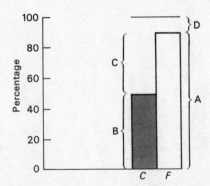

Fig. 5

FIG. 5.5. Child–family food difference for one 'category' (oils). A = Percentage of families eating any oil (90 per cent; food available). B = Percentage of children eating any oil (50 per cent). C = A – B; percentage of children not eating oil when available to family (40 per cent; child–family food difference or underusage). D = Percentage of families not eating oil (10 per cent; food unavailable). C = Child; F = family.

The food categories eaten by rural Sierra Leone children, according to age group, are compared with their families in Fig. 5.6. Relatively few families consumed vegetable protein foods (e.g. legumes), DGLV (dark green leafy vegetables), and fruits. The difference between the unhatched columns (family consumption) and the hatched columns (consumption by young children) for each food category indicates 'child/family food underusage'. For example, in children aged 6–11 months such 'underusage' was high in all food categories except cereals and fruits.

Child–family comparisons are especially relevant to nutrition education. Food available to the family and subsequent child usage are two aspects of child feeding. Education of the mother to provide certain food items for the child, such as 'animal proteins' (protein-rich animal products) will fail if such foods are not readily available, either purchased or locally grown.

Variety scores

A child's diet containing several different categories of foods is more varied and, thus, inherently more nutritionally desirable than a diet limited to a few categories. For this reason, a 'variety score' was constructed, based on the number of food categories consumed. For example, a child was allocated a score of one point for each food category of food consumed (based on the list indicated earlier). The total score could, therefore, range from zero to a maximum of eight; the higher the score, the greater the variety. Food combinations, as indicated in the 'variety score', cannot be considered as mixes or 'multi-mixes', since such foods were not necessarily eaten at the same time during the preceding 24 hours.

In the Cameroons, there was a significantly higher prevalence of under-nutrition (based on underweight and arm wasting) in children with a low

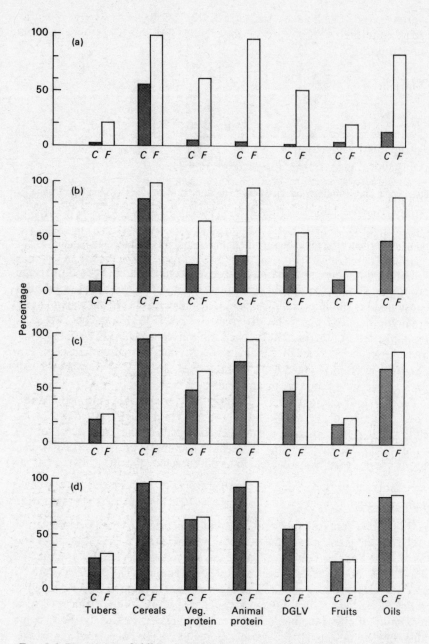

FIG. 5.6. Percentage of children and families consuming selected food categories (by age group of child) in rural areas of Sierra Leone. (a) Children aged 3–5 months; (b) those aged 6–11 months; (c) those aged 12–17 months; (d) those aged 18–30 months. *C* = child; *F* = family.

variety score (3 or under). In Sierra Leone, the relationships between low scores and high undernutrition prevalence were consistent only for acute undernutrition.

Mortality

General

Mothers reported the total number of their children born alive, the total still alive, and the number who died. From this was derived a crude 'child mortality index' (deaths/(live births + deaths/2)).

In Cameroon, even this crude index was discarded from the analysis, as the results were inconsistent with those from a more detailed national mortality study. Due to problems of recall, age-specific mortality rates were generally not attempted.

Results were used more for comparisons within each country than for national descriptions *per se*. These showed a higher mortality in the more rural areas, confirming the results of other studies.[71] In the two countries studied (Liberia and Sierra Leone), mortality rates of siblings were positively associated with chronic undernutrition in the sampled child. This suggests that stunting indicates the prior health experiences of siblings in the same families.

Measles

In Cameroon, mothers reported the number of children under 5 years of age in their family who had measles, the number who died from any cause, and those who died from measles during the preceding year. The overall CFR (Case Fatality Rate) for measles was 2.0 per cent in the two major cities, 4.1 per cent in other urban areas, and 6.2 per cent in rural areas, the last value being very close to the 7 per cent Morley found in Ilesha.[72] These results probably reflect the availability and use of health services, and general health and nutritional status as well as other factors, such as immunization. However, case fatality rates for children without measles were about the same throughout the country (ranging from 2.3 to 2.9 per cent).

For the whole country, measles contributed to an estimated 15 000 young child deaths (23.4 per cent of total deaths) during the past year. This was extremely persuasive evidence for the need to intensify immunization programmes throughout the country.

Morbidity

In the Cameroon survey, the mother reported the presence or absence of illness of the 3–24-month old over the preceding week. Such illness included being 'unwell' (unable to do 'usual' activities), diarrhoea (at least four watery stools a day), and fever, all for at least 2 days during the past week.

Mothers reported that approximately 30 per cent of these children had at least one type of illness, while 10 per cent had had all three. The prevalence of 'mild' wasting, based on a value under 85 per cent of the reference median weight-for-height, was 2–3 times higher in those children with reported 'illnesses'.

In the Liberia survey, teams measured rectal temperatures and examined for dehydration (marked loss of skin turgor, depressed anterior fontanelle) on a subsample of 224 young children aged 5–36 months. In 80 'normal' children (neither wasted nor stunted) throughout the nation, only 4 per cent had fever (over 100°F or 38°C) and none were clinically dehydrated. By contrast, among 144 children who were wasted or stunted, 15 per cent had fever and 19 per cent were dehydrated. These results are biased to some extent, because the surveyors knew the nutritional status of the sub-sampled children before recording temperature and dehydration. In addition, some of the temperature recordings were suspect. Despite these biases, the striking relationship between adverse nutritional status and current infections was apparent.

Quality control

Anthropometry

Quality control procedures were included for all anthropometric measures. Teams, supervisors, and senior survey personnel undertook pre-survey standardization exercises.[73-76] For each session, participants performed a single measure on 8–10 children or adults. Results compared all measures on the same child and were reviewed with teams shortly after each session. Senior personnel identified gross or consistent measurement differences and the reasons for the disparities demonstrated. Teams re-measured subjects under close supervision, as required.

Combined results for child measures showed an average difference (based on the square root of the mean square difference) among all teams of about 0.07 kg for weight, 0.2 cm for height, and 0.2 cm for arm circumference. Triceps fat-fold measures, particularly for mothers, were the least satisfactory. Checks by supervisors in the field suggested that the standard declined to some extent, although probably not seriously enough to compromise results.

The Salter hanging scale for weighing children and the insertion tape for measuring the arm circumference were portable and reliable.[39] When scales were checked using known weights, adjustments were minor and rare. All height–length boards were custom-made of wood (based on a model designed in UCLA) in the surveyed country. In Liberia, these boards were modified for use as containers for all team equipment. CDC have developed and marketed within the USA another height–length board, also of wood.

Other Measures

Quality control for other measures was limited to direct supervision of field workers by supervisors and senior investigators. Hence, no clear estimate of error has been possible. Lack of adequate time for probing for qualitative dietary recall by using more in-depth questions on intake, where required, may have resulted in underestimates of the numbers of subjects consuming certain foods, particularly oil, dark green leafy vegetables, and fruits. Special problems were determining all the foods contained in mixtures, soups, and sauces.

Haemoglobin

The collection of capillary blood in the field and its subsequent determination for haemoglobin levels was fraught with potential problems: improper collection techniques, inadequate protection of Drabkin's solution (ferricyanide) from the sun, and delay of 1–2 between collection and photometric reading in the laboratory, resulting in flocculation and other changes likely to invalidate results. Independent checks for two countries showed an agreement of haemoglobin within one gram % for 90 per cent of the estimations.

Conclusions

Practical outcomes

The real need for a national nutrition survey first needs to be resolved. Is there already sufficient evidence of regional differences throughout the country? What is the country's level of interest and will the survey results assist present or future programmes in improving health and nutritional status, including the development of useful low-cost surveillance system? What will be done with the results?

Basic information obtained

FNNS, using the collection of basic core information of weight, height, precise age, oedema and haemoglobin, have been able to identify the prevalence of PEM (protein-energy malnutrition) and anaemia in young children at one point in time in the nine countries (five in Africa) surveyed with UCLA or CDC collaboration. Most of the surveys have generated various subsequent nutritional activities, such as national seminars, or have been associated with the development of a system of surveillance.

Range of information

The use of FNNS can be increased by adding investigations, such as those pertaining to morbidity and mortality. However, careful and difficult decisions concerning cost, training, and time—both for data collection and analysis—need to be debated and agreed upon while the survey is being planned. The methodology requires flexibility to serve the needs of the country, and feasibility within financial, geographical, and climatic constraints.

The value of the results can be greatly increased when interpreted together with other nutrition-related information which may already be available from Government agencies in countries (e.g. food balance sheets, price and availability of major foods, etc.). Conversely, the efficiency and use of limited resources would be improved if a few selected practical indicators of nutritional status could be incorporated into agricultural, demographic, or other large-scale surveys when these are planned, as they are undertaken more often than nutritional surveys.

Country participation

Country participation in the survey (e.g. personnel and resources) was highly concerned with the field phase of FNNS and, even at that time, the UCLA representative played a major role. There is a danger of the methodology, although basic in concept, being purely imported without lasting benefits to the recipient country. Hence the methodology, if possible, should be modified appropriately to increase as much as possible the amount of planning, analysis, and report writing within the country surveyed, including the subsequent incorporation into surveillance or other continuing activities. At least one African country, Kenya, has conducted a nationwide rural survey, using the entire FNNS methodology with local resources.[64]

Survey design

Factors to be taken into account when planning any survey or FNNS include the source of population data, number of strata to be described, the sample size and methods, the subjects (such as age range of children) and the frequency, timing, and content of examinations. These surveys used recent census data and selected some 900 children aged 3–59 months from each stratum (thirty sites each of thirty children).

The multistage, cluster design (p. 62) was used to obtain random, representative samples, which were statistically valid and logistically attainable. In general this technique was satisfactory, given the constraints. However, in practice sampling was by no means ideal, having certain errors of non-response, incomplete young-child enumeration and, in one country (Lesotho), errors of non-coverage due to continuing inaccessibility resulting from heavy, early rains and a mountainous terrain. Improvement in sampling techniques perhaps would require major changes in the methodology or increase in cost. African countries have special logistical problems related to remote areas of road and transport limitations.

Interpretations of anthropometry

(a) Basic results. Anthropometry demonstrated that prevalence rates for chronic undernutrition varied much more within than among countries (Fig. 5.2). There was a consistent gradient within each country: lowest rates in the capital city, intermediate in urbanized, and highest in rural areas. Significant and practical differences in prevalence rates among administrative divisions (such as in the Cameroons) or between urban/rural zones suggested priority areas for nutritional improvements. Few children aged 3–5 months were undernourished, almost all being breast fed. All undernutrition rates rose after 6 months of age and reached a plateau in the second year of life. After that age, the prevalence of wasting became less whereas that for stunting usually remained about the same.

(b) Pros and cons of measures. Low height according to age (height-for-age) indicates chronic and low weight according to height (weight-for-height), which indicates acute PEM (protein-energy malnutrition). Each signify different durations of PEM. Chronic PEM suggests long-term social and nutritional problems—agricultural, diet, health and others—and acute PEM to short-term problems based at times on seasonal factors.

Because surveys were not conducted during the 'hungry' season due to logistic constraints, estimates for acute PEM (and oedema) prevalence were low. Results must be interpreted in this light or the alternative—a restricted FNNS repeated during representative seasons—should be considered, although it may be impractical.

Low weight-for-age ('underweight') and low arm-circumference-for-age ('arm wasting') cannot describe duration of PEM. However, the former is well known (e.g. growth charts) while the latter is convenient and relatively independent of precise age in young children.

(c) Reference levels and use of special groups. Analysis used 'international' reference data derived from US populations to convert body measures to specific indices of PEM. Index cut-off points designated the anthropemetric 'diagnosis' of PEM. For example, acute undernutrition was designated as less than 80 per cent of the reference level (median) weight-for-height. Such cut-off points are also of value in mortality predictions of young children.

Reference data can clarify the analysis and interpretation of measures by providing PEM prevalence rates. However, controversy continues about the value of 'international' reference data: the various overlapping sources of such data (Boston, NCHS/CDC, others); application to countries with different genetic backgrounds and socioeconomic status; the designation of cut-off points; the presentation of data (as percent of median, centiles, standard deviations).

The selection of local reference data for each country was not feasible. Instead, a special group (up to 100 for each year of age) of children of families with relatively high socioeconomic status was measured and results compared with the 'international' reference data. If there was little difference between the two, country nationals felt more comfortable with the international reference as akin to their own. In three of the four African countries, the genetic background of the special group was different from most of the country. However interpreted, reference data must never imply a standard of normality or excellence.

(d) Age. Proof of exact age is often impossible to obtain. This may compromise the value of age-dependent anthropometric indices such as height-for-age and weight-for-age. Such a bias can occur, especially when a high proportion of children have no written birth records. Precise age-independent indices, such as weight-for-height and arm circumference, may be more reliable measures. The latter, in particular, has the advantage that a tape is cheap and easy to use and carry.

Training and supervision

Training should emphasize the following: trainees from Government positions and with field experience; a manual appropriate for their use and oriented to tasks (especially practical anthropometric measurements, age determination, haemoglobin estimation, and sampling procedures); practice and review (with adequate evaluation methods) rather than theoretical aspects; field trials as close to the survey situation as possible.

Training included rigorous and repeated procedures for anthropometry. Formal 'standardization exercises' consisted of all teams and senior personnel measuring the same subjects. Results were satisfactory for weight, height, and arm circumference. Field supervision indicated that insufficient probing during dietary recall may have resulted in underestimates for certain foods, such as oils, dark green leafy vegetables, and fruits. Field collection of blood and the transport of the prepared solution was sometimes inadequate.

Interpretation of dietary data

Despite the serious, well-recognized limitations of the 24 hour dietary recall method, general patterns of young child feeding can be obtained. Such qualitative results, clearly expressed in graphic form (Fig. 5.4), show the extent and interplay of breast feeding, use of 'weaning food', 'family food', and 'any food' (as defined earlier) (p. 72).

These results can be valuable for comparative purposes—for example, to note any changes in breast feeding practices in the future. In addition, under-usage of foods available to the family diet can be expressed by the 'child/family food underusage' for a particular food category (Fig. 5.5) and by the 'variety score' (p. 75). Both can indicate important areas of emphasis for nutrition education.

References

1. Latham, M. C. 'Underlying causes of malnutrition and diagnosis of the nutrition situation'. In *Nutrition planning and policy for African countries* (edn. M. C. Latham and S. B. Westley). Cornell International Nutrition Monograph Series No. 5. Ithaca, New York, p. 8 (1977).
2. Cravioto, J., Birch, H. G., de Licardie, E. R., and Rosales, L. 'The ecology of infant weight gain in a preindustrial society'. *Acta paediatr. scand.* **56**, 71–84 (1967).
3. Mata, L. J. *The children of Santa Maria Cauqué: A prospective field study of health and growth.* MIT Press, Cambridge, Mass. (1978).
4. McGregor, I. A., Rahman, A. K., Thompson, B., Billewicz, W. Z., and Thomson, A. M. 'The growth of young children in a Gambian village'. *Trans. R. Soc. trop. Med. Hyg.* **62**, 341–52 (1968).
5. Morley, D. C., Bicknell, J., and Woodland, M. 'Factors influencing the growth and nutritional status of infants and young children in a Nigerian village'. *Trans. R. Soc. trop. Med. Hyg.* **62**, 164–95 (1968).
6. Jelliffe, D. B. *The assessment of the nutritional status of the community.* WHO, Geneva, monograph No. 53 (1966).
7. Schofield, S. *Development and the problem of village nutrition.* Institute of Development Studies, Sussex (1979).
8. Bailey, K. V. 'Malnutrition in the African region'. *WHO Chronicle* **29**, 354–64 (1975).
9. Interdepartmental Committee on Nutrition for National Defence. *Report of nutrition survey, Nigeria, 1965.* NIH, Bethesda, Md. (1967).
10. Government of Barbados. *The national food and nutrition survey of Barbados.* PAHO Scientific Publication, Washington, DC, No. 237 (1972).
11. Government of Guyana. *The national food and nutrition survey of Guyana.* PAHO Scientific Publication, Washington, DC, No. 323 (1976).
12. Jelliffe, D. B. and Jelliffe, E. F. P. 'The nutritional status of Haitian children'. *Acta trop.* **18**, 1–45 (1961).
13. Institute of Nutrition of Central America and Panama. *Nutritional evaluation of the population of Central America and Pamana. Regional summary.* US Departments of Health, Education and Welfare, Baltimore, Md. DHEW Publication No. (HSM) 72–8120 (1972).
14. Johnston, F. E. 'Identification and measurement'. In *Nutrition and malnutrition* (ed. A. F. Roche and F. Falkner). Advances in Experimental Medicine and Biology, Vol. 49. Plenum Press, New York (1974).
15. McKigney, J. 'Simplified field assessment of nutritional status'. Paper presented at *The XI International Congress of Nutrition, Rio de Janiero* (1978).
16. Miller, D., Nichaman, M. Z., and Lane, J. M. 'Simplified field assessment of nutritional status in early childhood: practical suggestions for developing countries'. *Bull. WHO* **55**, 79–90 (1977).
17. National Academy of Sciences. *Assessment of basic field surveys of nutritional status in young children.* Washington, NAS, DC, monograph No. 1 (1979).
18. Brink, E. W., Khan, I. H., Splitter, J. L., Staehling, N. W., Lane, J. M., and Nichaman, M. Z. 'Nepal nutrition status survey'. *Bull. WHO* **54**, 311–18 (1976).
19. Brink, E. W., Perera, W. D. A., Broske, S. P., Huff, N. R., Staehling, N. W., Lane, J. M., and Nichaman, M. Z. 'Sri Lanka nutrition status survey'. *Intl. J. Epidemiol.* **1**, 41–7 (1978).
20. Office of Nutrition. *Nepal nutrition status survey.* Mimeo report. DS/N, US AID, Washington DC, 20523 (1975).
21. Office of Nutrition. *Sri Lanka nutrition status survey* (1976).

22. Office of Nutrition. *Republic of Togo nutrition status survey* (1977).
23. Office of Nutrition. *Republic of Haiti national nutrition survey* (1978).
24. Office of Nutrition. *Arab Republic of Egypt national nutrition survey* (1978).
25. Office of Nutrition. *Liberia national nutrition survey* (1976).
26. Office of Nutrition. *Kingdom of Lesotho national nutrition survey* (1977).
27. Office of Nutrition. *Sierra Leone national nutrition survey* (1978).
28. Office of Nutrition. *United Republic of Cameroon national nutrition survey* (1978).
29. Blankhart, D. M. *Nutrition survey in Sierra Leone.* WHO, Brazzaville, document AFR/NUT/17 (1964).
30. Munoz, J. A. and Anderson, M. M. *Report on a nutrition survey in Basutoland.* WHO, Brazzaville, Document AFR/NUT/28 (1962).
31. May, J. M. 'The Federal Republic of Cameroon'. In *The ecology of malnutrition in West Africa.* Hafner, New York, pp. 205–48 (1970).
32. Behar, M. 'Appraisal of the nutritional status of population groups'. In *Nutrition in preventive medicine* (ed. G. H. Beaton and J. H. Bengoa). WHO, Geneva, p. 566 (1976).
33. Schaeffer, A. E. 'Assessment and monitoring of nutritional health'. Paper presented to *The XI International Congress of Nutrition, Rio de Janiero* (1978).
34. Ministry of Planning. *Household expenditure survey.* Central Statistics Office, Freetown, Sierra Leone (1977).
35. Ministry of Agriculture. *Statistical handbook.* Monrovia, Liberia (1976).
36. Serfling, R. E. and Sherman, I. L. *Attribute sampling methods for local health departments.* US DHEW, CDC, Atlanta, USA (1965).
37. Cochrane, W. G. *Sampling techniques,* 2nd edn. Wiley, New York, pp. 157, 247 (1963).
38. Cochran, W. G. and Cox, G. M. *Experimental design,* 2nd edn. Wiley, New York, p. 25 (1957).
39. Zerfas, A. J. 'Anthropometric field methods'. In *Human nutrition: a comprehensive treatise,* Vol. 2. Nutrition and growth (edn. D. B. Jelliffe and E. F. P. Jelliffe). Plenum Press, New York, pp. 348–62 (1979).
40. Nutrition Unit. *Measures of nutritional impact.* WHO, Geneva, Document WHO/FAP/79.1 (1979).
41. Waterlow, J. C., Buzina, R., Keller, W., Lane, J. M., Nichaman, M. Z., and Tanner, J. M. 'The presentation and use of height and weight data for comparing the nutritional status of groups of children under the age of 10 years'. *Bull. WHO* **55**, 489–98 (1977).
42. US Departments of Health, Education and Welfare. National Center for Health Statistics. *N.C.H.S. growth curves for children, birth—18 years.* DHEW, Baltimore, Md., DHEW publication No. (PHS) 78–1650 (1977).
43. Garn, S. and Clark, D. C. 'Trends in fatness and the origins of obesity, Ad hoc committee to review the Ten-State Survey'. *Pediatrics, Springfield* **57**, 455 (1976).
44. Tanner, J. M. Personal communication of data to CDC, USA (1977).
45. Neumann, C. G. 'Reference data'. In *Human nutrition: a comprehensive treatise,* Vol. 2 Nutrition and growth (ed. D. B. Jelliffe and E. F. P. Jelliffe). Plenum Press, New York, p. 300 (1979).
46. Goldstein, H. and Tanner, J. M. 'Ecological considerations in the creation and use of child growth standards'. *Lancet* **1**, 58 2–5 (1980).
47. Habicht, J. P., Martorell, R., Yarbrough, C., Malina, R. M., and Klein, R. E. 'Height and weight standards for pre-school children'. *Lancet* **1**, 611–15 (1974).
48. Tanner, J. M. 'Population differences in body size, shape and growth rate: a 1976 view'. *Archs Dis. Child.* **51**, 1–2 (1976).
49. Grant, K. and Seaman, J. 'Growth chart confusion' (letter). *Lancet* **3**, 102 (1979).

50. McLaren, D. S. and Burman, D. *Textbook of paediatric nutrition*. Churchill-Livingstone, London, pp. 106–12 (1976).

51. Keilmann, A. A. and McCord, C. 'Weight-for-age as an index of risk of death in children'. *Lancet* 1, 1247–50 (1978).

52. Chen, L. C., Chowdhury, A. K. M. A., and Huffman, S. 'Classification of energy-protein malnutrition by anthropometry and subsequent risk of mortality'. (Submitted for publication)

53. Waterlow, J. C. 'Classification and definition of protein calorie malnutrition'. *Br. med. J.* 3, 566–9 (1972).

54. Waterlow, J. C. 'Note on the assessment and classification of protein energy malnutrition in children'. *Lancet* 1, 87–9 (1973).

55. Hogan, R. C., Broske, S. P., Davis, J. P., Eckerson, D., Epler, G., Guyer, B. J., Kloth, T. J., Kolff, C. A., Ross, R., Rosenberg, R. L., Staehling, N. W., and Lane, J. M. 'Sahel nutrition surveys 1974 and 1975'. *Disasters* 1, 117 (1977).

56. Annegers, J. F. 'Seasonal food shortages in West Africa'. *Ecol. Food Nutr.* 2, 251–7 (1973).

57. Waterlow, J. C. 'Child malnutrition—the global problem'. *Proc. Nutr. Soc.* 38, 1–9 (1980).

58. Morley, D. 'National nutrition planning'. *Br. med. J.* 4, 85–8 (1970).

59. Jelliffe, D. B., and Jellifffe, E. F. P. 'Age independent anthropometry'. *Am. J. clin. Nutr.* 30, 242 (1971).

60. Zerfas, A. J. 'The insertion tape: a new circumference tape for use in nutritional assessment'. *Am. J. clin. Nutr.* 28, 782–7 (1975).

61. Jelliffe, D. B. and Jelliffe, E. F. P. 'The arm circumference as a public health index of PCM in childhood: current conclusions'. *J. prop. Pediatr.* 15, 253 (1969).

62. Shakir, A. and Morley, D. 'Measuring malnutrition'. *Lancet* 1, 758 (1974).

63. Pralhad Rao, N., Mathur, Y. C., and Chandra Harish. 'Age assessment of pre-school children by "local-event-calendar"'. *J. prop. Pediatr.* 21, 121 (1975).

64. Ministry of Finance and Planning. 'The rural Kenyan nutrition survey'. *Social Perspectives*. Central Bureau of Statistics, P.O. Box 30266, Nairobi, Kenya. Feb.–March (1977).

65. Habicht, J-P. 'Some characteristics of indicators of nutritional status for use in screening and surveillance'. *Am. J. clin. Nutr.* 33, 534 (1980).

66. Morley, D. *Paediatric priorities in the developing world*. Butterworths, London, p. 286 (1973).

67. Perez, C., Scrimshaw, N. S., and Munoz, J. A. Technique of endemic goitre surveys. In *Endemic goitre*. WHO, Geneva, Monograph series No. 44 (1960).

68. Pierson, R. *Report of a visit to Cameroon regarding vitamin A deficiency*. Mimeo. AID/Yaounde, Cameroon (1977).

69. Food and Nutrition Board. *Simplified dietary methodology*. Conference document, National Academy of Sciences, Washington, DC (1977).

70. Blankhart, D. M. 'Outline for a survey of the feeding and nutritional status of children under three years of age and their mothers'. *J. trop. Pediatr. Environ. Child Hlth.* 4, 176 (1971).

71. Morley, D. *Paediatric priorities in the developing world*. Butterworths, London, p. 211 (1973).

72. Ministry of Health. *Report by Medical Statistics Unit* (unpublished observations). Freetown, Sierra Leone (1978).

73. Habicht, J-P. 'Standardization procedures for the collection of weight and height data in the field'. In *Measurement of nutritional impact*. WHO, Geneva, WHO/ FAP/79.1 pp. 32–8 (1979).

74. Martorell, R. 'Identification and evaluation of measurement variability in the anthropometry of pre-school children'. *Am. J. phys. Anthropometry* **43**, 347 (1975).
75. Cameron, N. The methods of auxological anthropometry. In *Human growth*, Vol. 2 (ed. J. Falkner and J. M. Tanner). Plenum Press, New York, pp. 41–6 (1979).
76. Zerfas, A. J. Quality control in authropometry. Unpublished report (1974).

6 Nutritional rickets: a persisting problem

MEHARI GEBRE-MEDHIN

Rickets has been known to occur since ancient times. There are indications, from depiction of affected cases in early tombs, that the condition may have been recognized by the early Egyptians.[1] However, the disorder was not well known until about the middle of the seventeenth century. In western Europe, Wistler of Oxford is believed to be the first to have given a comprehensive description of the disease, in 1645, followed a few years later by Glisson,[2] in a report from Cambridge. Subsequently many contemporary practising physicians both in Europe and the Middle East gave vivid accounts of the disease, which was prevalent both in new industrial settlements, that were crowded and unhygienic, and in rural areas where strong local traditions and beliefs prevailed. Cod liver oil was used in parts of Europe as a traditional remedy against rickets and its nutritional value was accepted by some physicians of the time. Mellanby showed experimentally in 1918[3] that puppies with rickets responded to a fat-soluble vitamin. The vitamin was finally prepared in pure form in 1931. Treatment with this alone lasted until the end of the 1930s, until extensive investigations on the aetiology of rickets conclusively established the importance of a simultaneous lack of dietary vitamin D and sunlight in its development.[4]

Epidemiology and aetiology

Nutritional rickets is primarily attributed to an inadequate supply or availability of vitamin D. Under normal conditions the most important source of vitamin D is the conversion of 7-dehydrocholesterol in the skin to cholecalciferol by ultraviolet light. Cholecalciferol thus produced and that absorbed from food sources is metabolized in the liver and kidney to $1\alpha25$-dihydroxycholecalciferol, which is currently thought to be the major mediator of the biological actions of vitamin D.[5]

Particularly in temperate areas in the northern hemisphere, the supply of vitamin D from sunshine is uncertain, and at least during the winter months attention must be paid to the dietary supply. At the same time, it is important to remember that ordinary foods that have not been artificially enriched with vitamin D are generally very poor in their content of this vitamin and cannot be regarded as adequate substitutes for sunshine or be relied upon to supplement occasional sunning. It is therefore the currently held view that all children living in temperate areas should receive a concentrate of vitamin D.[2]

Rickets and exposure to sunshine

Until recently it was thought that rickets was rare—indeed non-existent— in warm, sunny areas of the world.[2,6] However, many investigations published during the last few years have emphatically shown that this view is incorrect.[2,7-10] An abundance of sunshine alone does not necessarily guarantee freedom from rickets. A combination of culturally conditioned aversion to sunlight, such as fear of the evil eye or a dark complexion, and poor housing that shuts out the rays of the sun, or socioeconomic conditions which make it necessary for the mother to work outside the home, means that many children may be kept indoors until they are able to crawl out into the sun. If infants are brought out they are usually completely and effectively covered up to avoid exposing them to the sun.

It has been speculated that skin pigmentation impairs the conversion of 7-dehydrocholesterol below the stratum corneum to vitamin D.[11] As a result of this, it is thought that under conditions of inadequate exposure to sunlight, children with a dark complexion will be more prone to develop rickets than those with less skin pigmentation. Exposure to sunlight is further reduced by the necessity to remain indoors or to wear extra clothing in cold winter weather. However, this possibility is not compatible with the variation observed in the incidence of rickets among different population groups with comparable skin pigmentation and living in areas with inadequate sunlight.[12]

A possible endogenous predisposition to rickets has been suggested by the familial or ethnic distribution of vitamin D deficiency observed in certain investigations.[13] However, the ready response of practically all cases of nutritional rickets to standard treatment with calciferol would seem to preclude the possibility that metabolic or genetic abnormalities play a major role in the aetiology of the disease.[8,12]

Relation to diet

The role of the calcium and phytate contents of the diet has been the subject of extensive investigations, but this issue remains controversial. Some authors have demonstrated satisfactory adaptation to lower levels of calcium intake and have found that prolonged ingestion of phytate is compatible with a normal development of bones and teeth.[14] Others have maintained that a habitually low intake of calcium does not appear to be deleterious to man, nor does an increase result in clinically detectable benefits.[15] By contrast, in a study of rural South African children with rickets showing no evidence of renal abnormalities, vitamin D deficiency, or the inherited varieties of rickets, Pottifor *et al.*[16] recently observed that a significant percentage of the children were hypocalcaemic and had increased serum alkaline phosphatase activities. Improvement was achieved with a normal diet in which the calcium content was increased by the addition of calcium supplements. A few cases have also been described of breast-fed premature infants in whom pure calcium deficiency in the presence of adequate vitamin D concentrations has been thought to cause a clinical picture of rickets.[17,18]

Rickets evidently does not have a single cause. A combination of culturally determined avoidance of sunlight, poor environmental conditions, and inadequate nutrition exacerbated by increased demands during periods of accelerated growth plays an important role.

Clinical diagnosis and pathophysiology

Risk factors

Rickets has its peak prevalence during the second half of the first year of life and usually does not appear much before the third month; it is seldom encountered after the third year (Table 6.1).[2,19] Premature infants are more prone to rickets than full-term babies.[20,21] Recently it has been suggested that pre-eclamptic toxaemia might be an aetiological factor in that not only is the transfer of calcium to the infant via the placenta affected[22,23] but that a degree of hypoxia and starvation attendant upon pre-eclampsia may adversely influence the vitamin D metabolism in the liver.

TABLE 6.1   *Age and sex distribution of rickets in Addis Ababa*[8]

Age	No. of infants		Total (%)
(months)	Male	Female	
0–5	60	27	87 (29.2)
6–11	85	46	131 (44.0)
12–23	40	28	68 (22.8)
24–35	8	4	12 (4.0)
Total	193	105	298 (100.0)

The artificially fed infant is more likely to develop rickets than the breast-fed baby. The reason for this is not fully known. There is still a great deal of controversy regarding the need for supplementary vitamin D in the breast-fed infant. Until recently it was thought that the vitamin D content of human milk is very low and is not increased by a high intake of the vitamin by the mother.[24] Some authors have even made forthright statements to the effect that the rare cases of rickets now encountered in certain areas are usually breast-fed infants whose mothers have failed to realize the need for vitamin D supplementation in their infants.[25-27] The demonstration of substantial amounts of aqueous-phase cholecalciferol sulphate in human milk[28] has given new fuel to the discussion of this issue. However, Leerbeck and Sondergaard[29] have recently reported that cows' milk contains 38 IU of vitamin D/1 and whole human milk 15 IU of vitamin D/1, and that 12 IU of the latter derives from the lipid fraction. These values are much lower than those reported previously in the literature. Furthermore, by means of chromatography they isolated cholecalciferol sulphate in an amount that was greater than that corresponding to the antirachitic activity of this substance. It remains uncertain whether the major part of the aqueous portion has any biological activity—indeed, whether the substance isolated is, in fact, vitamin D sulphate.[30]

Rickets is encountered more often in the winter months in temperate regions and in the rainy season in warm areas than during summer months. In a complete 12-month seasonal variation study in which they measured serum concentrations of 25-hydroxyvitamin D_3, Stryd *et al.*[31] found a strong correlation between the levels of this metabolite and the time of year, the approximate amount of sunlight, and temperature. The highest values for 25-hydroxyvitamin D_3 were recorded in the summer months and the lowest in the winter.

This study also showed a sex effect, the male values being slightly higher than the female ones, possibly reflecting behavioural differences between girls and boys or the effect of hormones on vitamin D metabolism in general.[32] Boys are more often affected by rickets than girls (Table 6.1).[8]

Clinical features

The clinical diagnosis of fully developed rickets seldom poses any problems for the physician who is aware of the possibility of encountering the disorder. The deficiency state is characterized by paleness, growth retardation, and delayed motor development, flabbiness of the muscles, fretfulness, and listlessness. Iron deficiency anaemia is often present, adding to the paleness. The abdomen is distended owing to the atony of the intestinal musculature and flatulence resulting from an excessively carbohydrate diet. Marked sweating is a common feature. The infant may be brought to a health centre because of coincidental respiratory or gastrointestinal upsets, which may conceal the rickets.

The skeletal manifestations of the disease are due to inadequate mineralization of osteoid at the epiphyses, resulting in soft bones that are liable to become deformed by normal physical activities. The earliest bone changes, called craniotabes, are usually observed over the parietal bones, where gentle pressure on unossified areas gives a brittle, snapping sensation. Other skeletal signs include enlargement of the epiphyseal cartilages of the long bones, which is particularly noticeable at three points: the costochondral junctions, producing rounded olive-shaped beading—the so-called rachitic rosary; at the lower ends of the radius and ulna, giving them a knob-like appearance; and at the lower ends of the tibia and fibula, causing characteristic widening of the malleoli, giving the impression of double malleoli on palpation.

If the disease continues into the second year of life, closure of the fontanelle and tooth eruption are delayed. Subsequently, formation of subperiosteal bone on the skull results in frontal and parietal bossing, giving the head a square or boxed appearance (caput quadratum). Later, weight-bearing and functional stress cause characteristic deformities. The forceful sucking in of the softened ribs on inspiration during respiratory infections produces bilateral depressions on the chest along the mid-axillary line. This deformity, in turn, gives the sternum undue prominence, causing the so-called 'pigeon breast' appearance. Sitting and crawling result in kyphosis of the spine, while standing causes scoliosis. When the child begins to walk, deformities of the legs add 'knock-knees' or 'bow-legs' to the clinical picture. Fractures of the extremities are not uncommon and occur at an early stage. The pelvic bones are also affected and it has been suggested that this may lead to difficulties in childbirth later on in life.

Radiological examination of the bones, usually the radius and ulna, shows a characteristic loss of definition of the metaphyseal line, with flaring and cupping of the bone ends, and generalized osteoporosis. Metaphyseal bands represent the healing stage.

Chemical pathology

In the presence of active rickets the serum phosphate concentration is generally lowered, the calcium concentration normal or slightly reduced, and alkaline phosphatase activity increased. Until recently, increased serum alkaline phosphatase activity used to be regarded as a constant feature. However, experience in India[33] and Ethiopia,[8] where rickets is commonly associated with protein-energy malnutrition, has thrown doubt on the consistent validity of this enzyme as a criterion of the vitamin D status. In a significant proportion of malnourished infants with rickets the serum alkaline phosphatase activity falls within the normal range.

Subclinical or early rickets may be difficult to recognize. Some degree of physiological craniotabes can often be observed in healthy breast-fed infants, and all infants, especially lean ones, show slight beading at the costochondral junction. It is also fairly common to find a hint of double malleoli in healthy children 5–6 months old. Uncertainty in evaluating these physiological findings may sometimes have led to overdiagnosis of rickets in community surveys.

Geographical distribution and resurgence of rickets

It has been pointed out above that until recently rickets was regarded as a disease of temperate areas that have only short periods of sunshine. Before the nineteenth century up to a third of young children were affected by the disease as a result of the growth of industrial centers to which people migrated in large numbers from rural areas.

During the early part of this century the incidence of rickets declined in Europe and North America as a result of the increasing use of cod liver oil as a supplement to children. The incidence of the disease was further decreased by the routine day-to-day vitamin D prophylaxis provided through antenatal and welfare clinics and supplementation of various foods with vitamin D.[2,6]

Available statistics indicate that the incidence of clinical rickets in industrialized countries in recent years has been very low. In the US prevalence figures of between 0.1 and 0.4 per cent of cases have been reported from teaching hospitals as well as in preschool nutrition surveys.[34,35] The disease virtually disappeared from Britain after milk fortification with vitamin D began in 1945.[2] In Sweden rickets was no longer commonly encountered after the late 1960s, when vitamin D supplementation of industrial food products was started.[36]

From about the mid-1940s, various reports began to show that rickets was very common in Third-World countries in the subtropics and tropics.[6] Data collected at hospitals and in community nutrition surveys in the eastern Mediterranian, the Indian subcontinent, south-east Asia, China, tropical Africa, and south America firmly established the high prevalence of the disorder in these regions.

A review of the situation during the past decade or two clearly indicates that rickets is still a regular feature of child morbidity in many developing countries[7,37] and severe cases, which are no longer encountered in industrialized nations, have recently been reported in these areas.[38]

Industrialized countries

Several reports published during the last few years strongly suggest that rickets may still be a significant problem in certain population groups, even in industrialized countries. It is striking that a diagnosis of rickets has been made fortuitously in children who were receiving regular paediatric supervision.[39] This may reflect a lack of awareness of the condition among the medical profession and illustrates the ease with which the clinician can be lulled into a false sense of security.

In the USA, the disorder has begun to reappear among children fed an essentially vegan diet, and a strong association with unsupplemented breast feeding has been suggested.[40] There seems to be a clear proponderance of coloured infants among these cases.[25,26] The children usually present with florid rickets characterized by the classical signs of the deficiency state, including seizures and pathological fractures. The dietary history reveals exclusion of the main sources of vitamin D and a lack of consistent supplemental vitamins. Some of the cases have represented minority groups living in densely populated neighbourhoods with heavy atmospheric pollution. Many of the families were advocates of a return to a more 'natural' diet free of food additives.

A resurgence of nutritional rickets has also been observed in Britain since the fortification of milk with vitamin D was stopped in the early 1960s. Now the disease is found mostly among Asian immigrants to the British Isles.[12] Some studies have yielded prevalence figures for biochemical rickets of up to 50 per cent in Asians and a prevalence of clinical rickets of 5 per cent.[12,41] The first two cases of neonatal rickets in the offspring of Asian immigrant mothers with osteomalacia were reported in 1973.[42] The extent of the rickets problem among this population group is further underlined by the necessity of carrying out several osteotomies for severe rachitic deformity.

Kourim and Skovrankova[43] report that nutritional rickets still occurs in central Europe. In a study of the incidence of this disorder in two Prague districts, the authors reviewed 193 cases between 12 and 24 months of age who were hospitalized for various reasons. Of these children, 5.2 per cent were found to have clinical, radiological, and biochemical signs of rickets, while nearly 30 per cent displayed an isolated increase in alkaline phosphatase activity. Inadequate ultraviolet irradiation due to air pollution and irregular administration of vitamin D were given as causative factors.

Less developed countries

Although the paradoxical occurrence of nutritional rickets in the warm and sunny areas of north Africa has been recognized for at least a decade and a half, the prevalence of the disorder remains excessively high. In a survey of children 0–36 months of age, Tabbane *et al.* found that 55 per cent of the studied population in Tunisia showed radiological signs of rickets.[19] The incidence of neonatal hypocaloaemia and infantile rickets is believed to be high among the Beduins in Israel,[4] but survey figures are seemingly lacking. Rickets in children under the age of 2 years is recognized as a problem in South Africa[45] and sporadic cases of the disorder in children over the age of 2 years have been seen for many years.[16]

Treatment and prevention

Health education

For most people of the world who are settled in areas where sunshine is plentiful, renewed energetic education in health and nutrition with regard to the necessity, efficacy, and safety of sunshine for infants and children, as well as for adult women, remains the pillar of rickets prevention and treatment. It must be considered unacceptable that rickets should occur in subtropical and tropical areas in the 1980s despite considerable sunning that should ensure more than a sufficient degree of endogenous cholecalciferol synthesis in the infants to protect against a deficiency state.

Although systematic information is lacking, experience in Ethiopia seems to indicate that the prevalence of rickets in highland parts of the country is now decidedly lower than it was in the late 1960s, when most surveys were carried out. This is probably the result of intensive campaigns carried out by the Ethiopian Nutrition Institute and the Ethio-Swedish Pediatric Clinic in Addis Ababa, including traditional organizations such as the Church. The information that was broadcast encouraged mothers to take their children out into the sun for a short time in the morning—the time of day when highland inhabitants intuitively and actively seek the warmth of ultraviolet light. The concept 'morning sun' was launched, implying, as it were inadvertently, that the 'immature' rays of the rising sun were unharmful, be it for the general complexion ('dark skin') or the face (point of entry for the 'evil eye') of the infant. It was also made clear that traditionally 'critical' parts of the body such as the face could be covered to avoid direct sunning. In due course this approach was found to be culturally acceptable to a population which is usually skeptical of modern health education and economically practical, since the need for purchasing pharmaceutical vitamin D supplement is avoided.

Knowledge of the ultraviolet content of sunshine in different geographical areas and the dose–response relation between exposure to sunshine and the rate of synthesis of 7- dehydrocholesterol, as well as of the transformation of the pro-hormone to its active 25-hydroxyderivatives, is still incomplete. In a study of severely rachitic Ethiopian twins, clinical, biochemical, and radiological signs of healing were observed after exposure to morning sun for about 30 minutes daily for a period of 3 weeks.[46] This amounts to a total exposure of 12 hours.

Changes in plasma 25-hydroxyvitamin D during long-term solar irradiation or short-term ultraviolet therapy and during oral treatment with 25-hydroxycholecalciferol or with vitamin D have been investigated. Ten times more vitamin D than the active derivative 25-hydroxycholecalciferol was required to produce equivalent plasma 25-hydroxyvitamin D concentrations. Changes in plasma 25-hydroxyvitamin D concentrations during solar irradiation were found to be equivalent to those produced by a daily dose of 10 000 IU of oral vitamin D, which is 100 times the recommended daily adult dietary intake,[47] but the authors are cautious in accepting this figure as representing endogenous synthesis. However, an interesting aspect of vitamin D metabolism is implied by these observations. It seems that synthesis of vitamin D in such generous amounts during summer months, coupled with pronounced but hitherto poorly understood delays in the transformation of the vitamin to its active 25-hydroxy derivatives, may provide some protection against deficiency in winter.

Vitamin D supplements

Cod liver oil provides approximately 300 IU of vitamin D per teaspoon (5 ml) and can cure uncomplicated infantile rickets. However, large therapeutic doses (1000–5000 IU) are desirable and give a more rapid recovery. Severely affected children require such high daily doses for approximately 2 weeks, followed by prophylactic doses (400 IU) during the winter.

In many parts of the world where rickets is highly prevalent, health services are poorly developed and routine supplementation of infants' and children's diets with vitamin D is not practised. Most of the infants in these areas are rarely seen by medical teams and those who seek emergency help are seen

once, and often not again for months. Therapeutic day-to-day administration of vitamin D is rendered difficult by a combination of environmental
and socioeconomic factors. In cases such as these a single massive dose of
vitamin D, the so-called 'stoss' method of Harnapp,[48] can be administered. It
is well known that vitamin D, even when given in a dose of 250 000–500 000
IU, is well absorbed and stored in the liver and liberated for use as required.[49]
Massive dose vitamin D therapy has previously been used in industrialized
countries, but its practice has been discouraged because of serious toxicity
problems. However, in recent years this method of treatment has been given
new consideration in situations where a significant percentage of infants are
at risk of developing rickets and where the addition of vitamin D to foods is
not practised. Interestingly, the effect of oral vitamin D is more rapid than of
intramuscular medication.[8] Further, single massive dose vitamin D therapy
seems effective under conditions in developing countries with an
accumulation of infections, including diarrhoeal disease. Even in severe cases
of malnutrition the capability of converting vitamin D into its active metabo-
lites does not seem to be lost.[50] Administration of vitamin D orally by 'stoss
dosage' has been found to be safe and effective in large-scale trials in
Ethiopia[8] and Morocco.[51]

Conclusion

It has been repeatedly stated above that the previous occurrence and current
resurgence of nutritional rickets are the result of the combined effects of a
number of causative factors, including insufficient sunshine, poor housing
conditions, and inadequate nutrition. It is unlikely, therefore, that isolated
preventive or therapeutic efforts can provide an answer to the problem.
Ultimate control over the disorder will be accomplished only when the
benefits of education, an increased standard of living, and expanded health
services reach the average citizen.

In industrialized countries, special attention needs to be given to children
at risk, notably twins, premature births, and offspring of vegans. In Britain,
where rickets is a problem largely in Asian immigrants, a recent 1980
Working Party[52] has advised against fortification of food with vitamin D
because of the risk of hypercalcaemia. Instead, it recommended a health
education approach coupled with individual vitamin D supplementation
where required.

References

1. Sabri, I. A. *J. R. Egypt. med. Assoc.* **26**, 166 (1943).
2. Davidson, S., Passmore, R., and Brock, J. F. *Human nutrition and dietetics.*
   Churchill, Edinburgh (1972).
3. Mellanby, E. *J. Physiol. (Lond.)* **52**, xi, liii (1918).
4. Chick, H., Macral, T. F., Martin, A. J. *et al. Biochem. J.* **32**, 12 (1938).
5. Omdahl, J. L. and DeLuca, H. F. Regulation of vitamin D metabolism and function. *Physiol. Rev.* **53**, 327 (1973).

6. Jelliffe, D. B. *Infant nutrition in the subtropics and tropics.* WHO, Geneva, monograph series No. 29 (1967).
7. Aust-Kettis, A., Björnesjö, K. B., Mannheimer, E. *et al.* Rickets in Ethiopia. *Ethiopian Med. J.* **3**, 109 (1965).
8. Gebre-Medhin, M. and Vahlquist, B. Effect of single massive dose vitamin D therapy (oral or intramuscular) on rickets in Addis Ababa children. *Courrier* **22**, 12 (1972).
9. Salimpour, R. Rickets in Tehran. *Archs Dis. Child* **50**, 63 (1975).
10. Galal, O. M., El-Nabawy, M., and Hassan, A. Incidence of rickets in two children populations in Egypt. *Aim Shams Med. J.* **21**, 133 (1970).
11. Loomis, W. F. Skin-pigment regulation of vitamin-D biosynthesis in man. *Science* **157**, 501 (1967).
12. Goel, K. M., Logan, R. W., Arneil, G. C. *et al.* Florid and subclincial rickets among immigrant children in Glasgow. *Lancet* **1**, 1141 (1976).
13. Doxiadis, S., Angelis, C., Karatzas, P. *et al.* Genetic aspects of nutritional rickets. *Archs Dis. Child* **51**, 83 (1976).
14. Walker, A. R. P. Serials, phytic acid and calcification. *Lancet* **2**, 244 (1951).
15. Walker, A. R. P. The human requirement of calcium: should low intakes be supplemented? *Am. J. clin. Nutr.* **25**, 518 (1972).
16. Pottifor, J. M., Paddy, R., Wang, J. *et al.* Rickets in children of rural origin in South Africa: Is low dietary calcium a factor? *J. Pediatr.* **92**, 320 (1978).
17. Kooh, S. W., Fraser, D., Reilly, B. J. *et al.* Rickets due to calcium deficiency. *New Engl. J. Med.* **297**, 1265 (1977).
18. Maltz, H. E., Fish, M. B., and Holliday, M. A. Calcium deficiency rickets and the renal response to calcium infusion. *Pediatrics* **46**, 865 (1970).
19. Tabbane, C., Bousnina, S., and Karoui, M. K. Actualité du rachitisme à Tunis. Resultats d'une enquête intrahospitalière. *Rev. Tun. Péd.* **1**, 1 (1979).
20. von Sydow, G. V. A. Study of the development of rickets in premature infants. *Acta paediatr.* **33**, Suppl. 2, 122 (1964).
21. Griscom, N. T., Craig, J. N., and Neuhasuer, E. B. D. Systemic bone disease developing in small premature infants. *Pediatrics* **48**, 883 (1971).
22. Bosley, A. R. J., Verrier-Jones, E. R., and Campbell, M. J. Aetiological factors in rickets of prematurity. *Archs Dis. Child* **55**, 683 (1980).
23. Khattab, A. K. and Forfar, J. O. The interrelationship between calcium, phosphorus, and glucose levels in mothers and infants in conditions commonly associated with 'placental insufficiency'. *Biol. Neonate* **18**, 1 (1971).
24. Harris, R. S., and Bunker, J. W. M. Vitamin D potency of human breast milk. *Am. J. Public Hlth.* **29**, 744 (1939).
25. O'Connor, P. Vitamin D-deficiency rickets in two breast-fed infants who were not receiving vitamin D supplementation. *Clin. Pediatr.* **16**, 361 (1977).
26. Edidin, D. V., Levitsky, L. L., Schey, W. *et al.* Resurgence of nutritional rickets associated with breast-feeding and special dietary practices. *Pediatrics* **65**, 232 (1980).
27. Bachrach, S., Fisher, J., and Parks, J. S. An outbreak of vitamin D deficiency rickets in a susceptible population. *Pediatrics* **64**, 871 (1979).
28. Lakdawala, D. R. and Widdowson, E. M. Vitamin-D in human milk. *Lancet* **1**, 167 (1977).
29. Leerbeck, E. and Söndergaard, H. The content of vitamin D in human milk and cows' milk. *Br. J. Nutr.* **44**, 7 (1980).
30. Hollis, B. W., Lambert, P. W., and Draper, H. H. Abstracts of papers for the third joint meeting of the American Institute of Nutrition, the American Society for Clinical Nutrition and the Nutrition Society of Canada. *J. Nutr.* **109**, xxiv (1979).

31. Stryd, R. P., Gilbertson, T. J., and Brunden, M. N. A season variation study of 25-hydroxyvitamin D_3 serum levels in normal humans. *J. clin. Endocrinol. Metab.* **48**, 771 (1979).
32. Castillo, L., Tanaka, Y., DeLuca, H. F. *et al.* The stimulation of 25-hydroxyvitamin D_3-1α-hydroxylase by estrogen. *Archs Biochem. Biophys.* **179**, 211 (1977).
33. Reddy, V. and Srikantia, S. G. Serum alkaline phosphatase in malnourished children with rickets. *J. Pediatr.* **71**, 595 (1967).
34. American Academy of Pediatrics Committee on Nutrition. Infantile scurvy and nutritional rickets in the United States. *Pediatrics* **29**, 646 (1962).
35. Owen, G. M., Kram., Garry, P. *et al.* A study of nutritional status of preschool children in the United States 1968–1970. *Pediatrics* **53** Suppl., 597 (1974).
36. Sjölin, S. How should the vitamin D requirement of the infant be met? (In Swedish.) *Nordisk Medicin* **94**, 215 (1979).
37. World Health Organisation. *Joint FAO/WHO Expert Committee on Nutrition.* WHO Technical Report Series No. 377 (1967).
38. Wolde-Mariam, T. and Sterky, G. Severe rickets in infancy and childhood in Ethiopia. *J. Pediatr.* **82**, 876 (1973).
39. Rudolf, M., Arulanantham, K., and Greenstein, R. M. Unsuspected nutritional rickets. *Pediatrics* **66**, 72 (1980).
40. Dwyer, J. T., Dietz, W. H., Hass, G. *et al.* Risk of nutritional rickets among vegetarian children. *Am. J. Dis. Child.* **133**, 134 (1979).
41. Ford, J. A., Colhoun, E. M., McIntosh, W. B. *et al.* Rickets and osteomalacia in the Glasgow Pakistani community, 1961–1971. *Br. med. J.* **2**, 677 (1972).
42. Ford, J. A., Davidson, D. C., McIntosh, W. B. *et al.* Neonatal rickets in Asian immigrant population. *Br. med. J.* **3**, 211 (1973).
43. Kourim, J. and Skovrankova, J. On the incidence of rickets in two Prague districts. *Cs. Pediatr.* **34**, 42 (1979).
44. Biale, Y., Shnay, S., Levi, M. *et al.* 25-hydroxycholecalciferol levels in Beduin women in labor and in cord blood of their infants. *Am. J. clin. Nutr.* **32**, 2380 (1979).
45. Dancaster, C. P. and Jackson, W. P. U. Studies in rickets in the Cape Peninsular. *S. Afr. Med. J.* **35**, 890 (1961).
46. Höjer, B. and Gebre-Medhin, M. Rickets and exposure to sunshine. *J. trop. Pediatr. Envir. Child Hlth.* **21**, 88 (1975).
47. Stamp, T. C. B. and Haddad, J. G. Comparison of oral 25-hydroxycholecalciferol, vitamin D, and ultraviolet light as determinants of circulating 25-hydroxyvitamin D. *Lancet* **1**, 1341 (1971).
48. Harnapp, G. O. *Klin. Sochenschr.* **17**, 390 (1938).
49. Windorfer, A. Über die Vitamin D Resorption bei Verabreicherung hoher Dosen (Vitamin D-Stoss). *Klin. Wochenschr.* **17**, 228 (1938).
50. Höjer, B., Gebre-Medhin, M., Sterky, G. *et al.* Combined vitamin-D deficiency rickets and protein-energy malnutrition in Ethiopian infants. *Envir. Child Hlth.* **23**, 73 (1977).
51. Guinard, J., Belkziz, O., and Belhaj, A. Assay on the systematic prophylaxis against rickets in public health practice in Morocco. [In French.] *Bull. Acad. Natl. Med.* **155**, 662 (1971).
52. Working Party on Fortification of Food with Vitamin D. Committee on Medical Aspects of food Policy 1980. *Report on Health in Society*, No. 19. HMSO, London (1980).

7 Bladder-stone disease in children

AREE VALYASEVI

Introduction

This disease may be called idiopathic vesical lithiasis; it is a term used to indicate a condition in which there is an absence of any known local predisposing cause in the bladder itself or of such other factors as endocrine or metabolic disorders, primary infection, prolonged restrict mobilization, or foreign body. Bladder-stone disease is known to have occurred from the time of the earliest recorded history of man. Egyptian mummies, dated as far back as 4800 BC, showed indications of stone formation.[1] It is not possible to document the entire history of bladder-stone disease from the ancient times; however, the vast majority of stones are present in young males who do not have concurrent renal calculi, and after surgical removal recurrences are rare. The older medical literature of the United States, western Europe, Great Britain, and elsewhere contains many references to the problem of bladder-stone disease, particularly in young children, before the nineteenth century. Since then, this disease has gradually disappeared in the developed countries.

During the twentieth century, bladder stones in children are or recently have been clinically important in several countries in the middle-east, far-east and south-east Asia. These include Turkey, Egypt, Syria, Tunisia, Iran, Pakistan, India, Burma, Thailand, Laos, and Indonesia.[2] The geographical distribution of bladder-stone disease has changed with the passage of time, resulting in a disappearance of the disease from the more industrialized and affluent societies to rural areas of agricultural developing countries. It is interesting to note, however, that vesical lithiasis has not been reported to occur with high incidence in children in south America, central or south Africa.

Natural history and clinical manifestations

Bladder stone is either an acute or a chronic disease. Onset may be sudden, with acute urinary tract obstruction in the absence of preceding symptoms. More often patients or their parents note the onset of cloudy urine which leaves a white, sandy residue when dried. After a period of weeks or months patients may note interruption of the urinary stream, with bladder or urethral pain at the end of voiding. There is progressive hesitation, dysuria, and frequency, and—in the child with obstructive disease—marked agitation. Patients may experience rectal prolapse or rupture of conjunctival vessels while straining to force urine passed through a 'ball valve' obstruction. Voiding may be accomplished only in a recumbent position. In males, penile hypertrophy may result from priapism and pulling in an attempt to start urine flow. Haematuria, seldom severe, may occur after strenuous efforts to void. About one-quarter of patients admitted to Ubol Hospital, north-east Thailand, had a history of spontaneous stone passing before operation. In the endemic stone area, diagnosis was made mainly by patient's history and urethral sound; radiological examination was seldom used.

Stones are of different size, shape, and colour. The largest stone in the collection at Ubol Hospital, north-east Thailand, weighing 1000 g, was removed from a 60-year-old man who had symptoms for about 40 years. Some stones are flat, some triangular, and some shaped like 'collar buttons' or dumb-bells. Most are oval and smooth. When more than one stone is present, highly polished facets are seen. The stone colour varies from chalky to creamy white and from tan to dark brown. Analysis of Thai bladder stones has revealed mixtures of various stone-forming salts and are often composed of ammonium acid urate, uric acid, and calcium oxalate.[3] These analyses are similar to observations on the composition of vesical calculi made by several authors during a period of high prevalence of the disease in Europe in the nineteenth century.

In those areas where the disease is endemic in children, some similarities can be seen as follows:

1 It occurs predominantly in children under 10 years of age.
2 The rate of hospital admissions is higher for boys than for girls.
3 Renal stones are not usually associated with the disease.
4 The rate of recurrence after surgical removal is low in comparison with renal lithiasis in adults.
5 The stones are usually composed of calcium oxalate and ammonium acid urate.
6 The disease generally occurs among children of low economic class. However, it has not been associated with overt signs of vitamin deficiencies or protein-energy malnutrition.

The natural history and clinical manifestations are quite different from stones of the upper urinary tract, which occur mostly in adults all over the world. These observations suggested that stones of the upper urinary tract are likely to differ from the bladder-stone disease in aetiological factor(s); however, the mechanism of stone formation may be the same. Therefore, information gained from the bladder-stone studies can be applied in the preventive measures of this disease as well as providing a better understanding of the stone formation in the upper urinary tract.

Hospital incidence

The hospital admissions for bladder-stone disease in different countries reported at the WHO Regional Symposium on Vesical Calculus, held in

Bangkok in 1972,[4] are shown in Table 7.1, and continued to be reported among these countries during the Second Symposium held at the Fogarty International Center, Bethesda, Md., USA in 1976.[2] However, there are no data available on the prevalence rate in general population except in Thailand, where medical records were reviewed and indicated a seriousness of the problem. Unakul,[5] reported that during the period of 1953–59 there were 26 101 patients treated in about 70 Government hospitals in Thailand for urinary tract stones. Among these, 82 per cent were bladder stones found predominantly in children of the north and north-east regions.

TABLE 7.1   *Hospital admissions for bladder-stone disease in different countries*

Reference	Country	Years	No. of patients admitted
1. Ramli, H.	Indonesia	1950–54	871
Dr. Tjipto, M. Hospital	Djakarta	1966–70	549
2. Sadre, M.	Iran	1968–71	166
Teheran University Hospital			
3. Loutfi, A.	Egypt	1951–70	619
Cairo Children's Hospital			
4. Benjamin V.	India		
	Vellore	1959–69	248
	New Delhi	1955–62	116
5. Chutikorn, C.	Thailand	1956–62	4271

Admissions for bladder stones peaked from the age 3–4 years, while 47 per cent of all admissions were children under the age of 5 years (Fig. 7.1). Ureteral and kidney stones had an entirely different age distribution; 96 per cent of stones in these sites were found in patients of 20 years or older. Urethral and bladder stones occurred more predominantly in males (10 : 1); whereas less prominent ratio (4 : 1) was found in ureteral and renal stones.[6]

Epidemiology

Since hospital statistics in Thailand had shown a high incidence of the disease in the north and north-east, and suggested significant differences in the pre- valence in subpopulations of Ubol Province in the north-east, an epidemi- ological study was carried out in greater detail in this province.[7] Forty-four villages and three towns were surveyed by interviewing 20 806 people. The study was designed to establish the prevalence rate of stone hospitalization, stone symptoms, and its associated factors in more detail. It was found that 253 (1.2 per cent) had a history of stone passing or stone operation; an addi- tional 540 (2.6 per cent) had had one or more of the following symptoms: difficulty in urination, sandy or cloudy urine, haematuria or painful urin- ation (classified as presumptive symptoms of bladder stones). For every

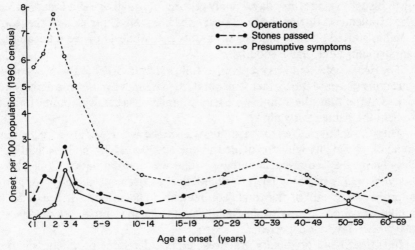

FIG. 7.1. Age of onset of bladder-stone episodes in sample population, Ubol province, 1963.

twelve individuals who encounter bladder-stone disease, one was hospitalized for removal of bladder stone, three had spontaneous stone passage, and eight had presumptive symptoms. In the sampled village population, the life-long prevalence rate of positive stone episodes was 1420 per 100 000; prevalence rates were significantly lower among residents of towns with populations of 5000 or more, the lowest being 210 per 100 000 in the Ubol Municipality (population 27 000).

Age-specific rates for onset of symptom suggested that there could be two periods of the genesis of primary urinary stones: early infancy and early adulthood. In children, the disease occurred before 1 year of age and reached the highest peak at the age of 3 years. Females comprised only 9 per cent of bladder-stone patients admitted to Ubol hospital, 18 per cent of persons had passed stones but did not require an operation, and 32 per cent of those had presumptive symptoms of the disease. These data suggested that females, possibly for anatomical reasons, developed surgically mature stones less often than did males.

Diet and nutrition

The difference in the prevalence rates between rural (high) and urban (low) areas within Ubol province has led to a comparative study of infant feeding practices, dietary habits and intakes, and urinary constituents among the inhabitants of the village of Nong Kohn and the town of Ubol.[8-10] These two sites are only 16 km apart and the pertinent findings are summarized below.

Early introduction of rice feedings: about 60 per cent of the village families started their infants on glutinous rice feedings during the first week after birth, usually on the third day. Ninety per cent of the village children received rice supplements during the first 4 weeks of life. Only 3 per cent of the town families started on rice supplements at this age, while 47 per cent began rice supplements later than 3 months.

The rice supplement was prepared as follows: the mother chews the cooked glutinous rice until it becomes semi-solid, then wrapped in banana leaves and baked. After that, the infants were fed by hand with a small amount of water to help the babies to swallow.

The amount of rice fed to the infants was on the average of 50–55 g per day (uncooked weight) which provided about 60–70 kcal/kg body weight per day. The energy requirement of these infants was approximately 120 kcal/kg body weight per day. Therefore the rice supplement already supplied approximately half of the total requirements. The amount of breast milk intake was subsequently studied and found to be low, ranging from 400 to 600 ml per day,[11] as expected.

In further work on dietary practices and bladder-stone prevalences in the north, north-east (endemic stone area) and central (non-endemic) Thailand, a striking correlation was found between early introduction of rice and high prevalence of the disease.[12] Infants who were fed on rice during the first week of life had a prevalence rate of bladder-stone disease twice as high as those who were started on rice later in life. However, families who consumed glutinous rice more often fed rice to neonates than did families eating ordinary rice.

It is also of special interest to note that, in England (Norfolk County), during the period 1772–1828, it was postulated that the incidence of bladder stone in earlier times was related to the frequency of the use of non-milk foods for 'hand-feeding' infants. This factor could account for the high incidence of bladder stones in cities and in non-dairying central and south England.[13]

Dietary sources of oxalate

Previous dietary survey data revealed the consumption of varieties of vegetables and leaves of forest plants among the village inhabitants.[8,12] Determination of the oxalate contents of these vegetables and plant leaves were carried out and found to be very high, ranging from 1589 to 4858 mg/100 g dry weight,[14] as shown in Table 7.2. Among these, tampala and bamboo shoots are most often given to the infants and young children. It has been estimated that the usual amount of oxalate consumption significantly increased the oxalcrystalluria of all crystal sizes and the clumping.[15] Further increase in oxalcrystalluria of medium- and large-sized crystals with clumping was demonstrated when the subjects consumed more of the vegetables of this kind.

TABLE 7.2   *Oxalic acid content in vegetables commonly consumed in Ubol villages*

Common name	Specific name	Oxalic acid (mg/100 g fresh weight)	Oxalic acid (mg/100 g dry weight)
Phak pel	*Polygonum odoratum* Lour.	956	4195
Cha phlu	*Piper sarmentosum* Roxb.	691	3888
Tampala*†	*Amaranthus gangeticus* L.	402	3349
E-Heen‡	*Monochoria vaginalis* Presl.	219	4858
Bamboo-shoot†‡	*Bambusa* spp	113	1589
Samek*	*Syzygium* spp	112	—

* Most commonly consumed by adult villagers (every day, all years).
† Also commonly consumed by infants.
‡ Most commonly consumed by adult villagers (every day, May–October).

Urinary excretions and composition of nutrients and metabolites

Information deriving from these urinary studies may: (a) reflect the adequacy of some nutrient intakes which are relevant to the bladder-stone disease; and (b) indicate the level of substances which relate to stone formation.

The studies in Thailand[9,10] revealed that the 24-hour urine volumes were often less in village newborn infants and children under 10 years old than in those of comparable age in the city. The small urine volumes, especially in the newborn period, suggested that some degree of dehydration existed in the village infants. The urinary osmolarities were generally lower in samples of the villagers than in those of the city dwellers, particularly in village newborns less than 15 days of age (6.5±2.0 *vs.* 16.2±5.2 mos/24 h urine). These low urinary osmolarities were due to low urinary sodium, chloride, and potassium concentrations, especially shown in infants over 6 months old. The urinary pH was about 6.0 in both groups of infants during the first 2 weeks of life, then gradually increased to 6.6 in the village subjects aged between 7 and 12 months.

Striking differences were found on the urinary phosphate and calcium excretion values. Significantly lower concentrations and 24-hour excretion of urinary phosphate was found in the rural infants than those of the city group, especially in the first year of life. On the contrary, the urinary calcium concentration and 24-hour excretion values were higher in the village than in the city samples, especially in the newborn age group. It appears that limited excretion of phosphate by village subjects is a reflection of marginal dietary intake. It was also found that urinary pyrophosphate was also significantly lower in the village subjects as compared with their city counterparts. Pyrophosphate will inhibit both hydroxyapatite and calcium oxalate precipitations; therefore, the low urinary pyrophosphate deriving from low dietary phosphate intake enhances the process of stone formation.

The urinary oxalic acid excretion values were significantly higher in village urine samples than the city or Bangkok counterparts (136 ± 12 vs. 99 ± 8 mg/g creatinine). This finding concurs with the dietary study, which indicated high consumption of dietary oxalate that was contained in local vegetables and plant leaves.

It should also be noted that the urinary phosphate excretion in bladder-stone patients admitted to the children's hospital in Cairo, Egypt (Loutifi 1977) and in Karachi, Pakistan (Rahman 1977) were also reported to be significantly low.[16]

Occurrence of oxalate and uric acid crystalluria

One of the most striking observations made in the series of studies in Thailand concerned with the occurrence of crystalluria in infants and children living in endemic-bladder-stone areas. Oxalate crystalluria was commonly observed in village urine samples from all age groups.[9,10] Forty-three per cent of the village infants under 45 days of age exhibited oxalcrystalluria alone or with uric acid crystalluria, whereas none of the city subjects of the same age group did. Furthermore, 46 per cent of the village infants under 12 months of age showed either only oxalate crystalluria or oxalate crystalluria with uric acid crystals; whereas only 20 per cent of the city subjects exhibited such crystals. In addition, village children 2–10 years of age had a higher incidence of calcium oxalate crystalluria than subjects of the same age living in Ubol city.

As previously described, dietary sources of oxalate in village infants and young children in endemic stone area can produce heavy oxalcrystalluria and crystal clumping. Therefore, it is likely that exogenous oxalate play an important role in the oxalcrystalluria commonly found in the urine of village subjects in north-east Thailand.

It is also of special interest to observe the relationship between maternal diet and crystalluria in their breast-fed infants. The study was carried out in twenty-five village mothers and their infants with various levels of oxalate consumption.[17] When the mothers consumed vegetables containing high oxalate concentrations urine of the mothers and their infants showed a considerable increase in the level of oxalcrystalluria within 48–72 hours. This finding indicates that oxalate in the maternal diet can be excreted in the breast milk which, in turn, induces oxalcrystalluria in young infants.

Comparison of the urinary calcium, oxalic acid, phosphorus, pyrophosphate, uric acid, ammonia, pH, and crystalluria between the samples containing only oxalate crystals and the other with mixed oxalate, uric acid, and ammonium urate crystals is shown in Table 7.3 The urine that showed only oxalcrystalluria had only mild to moderate degree of crystalluria. On the contrary, urine with mixed oxalate, uric acid, and ammonium urate crystals had a moderate to heavy degree of crystalluria and contained significantly higher concentrations of calcium as well as uric acid and ammonia. These findings will be further discussed when considering the mechanism of stone formation.

TABLE 7.3 *Occurrence of crystalluria and mean ± SE value of urinary calcium, oxalic acid, phosphorus, uric acid, ammonia, and pH*

		Oxalate crystal		Mixed oxalate uric acid and ammonium urate crystals
	No.	Mean±SE (mg/100 ml)	No.	Mean±SE (mg/100 ml)
Calcium	10	4±1	10	9±2**
Oxalic acid		5±1		5±1
Phosphorus		12±7		62±34****
Pyrophosphate		0.03±0.08		0.09±0.04
Uric acid		39±4		68±13***
Ammonia		43±9		96±15*
pH		6.5±0.2		6.1±0.2
		Crystalluria		
Calcium oxalate	9/10	Mild to moderate†	9/10	Moderate to heavy
Uric acid	0/10	—	6/10	,, ,,
Ammonium urate	0/10	—	5/10	Heavy

* P<0.01; ** 0.01<P<0.02; *** 0.02<P<0.05; **** 0.05<P<0.10.
† Mild = Few to nine crystals per high-power field; moderate = 10–29 crystals per high-power field; heavy = over 30 to loaded crystals per high-power field.

Possible role of orthophosphate in bladder-stone disease

As an aetiological factor

It has been demonstrated that the village infants and young children excreted remarkably low urinary phosphate, which is likely to be due to low dietary phosphate intake. Oral orthophosphate supplementation* was given to village subjects, ages ranging from 6 to 19 months, in addition to their regular feeding practices.[18,19] Microscopic examination of the urine revealed that, within 24 hours of the supplementation, oxalcrystalluria disappeared. The chemical analyses of the urine showed a marked increase in urinary total phosphate and pyrophosphate; at the same time, significant decreases in urinary calcium and oxalic acid excretions were also found. These changes may explain the disappearance of oxalcrystalluria. Furthermore, it can also be interpreted that infants and children living in the endemic stone area of north-east Thailand consume low dietary phosphate.

The above findings are of particular interest since weaning rats, when fed on low phosphate diet, produced high incidences of mild renal and bladder

* Administered as 60 mg phosphorous/kg body weight per day in 6.6 ml of solution containing 2.32 g Na_2HPO_4 and 0.36 g KH_2PO_4, with neutral pH.

citrate urolithiasis.[20-22] Calcium, phosphorus, and citrate interactions in oxalate urolithiasis formation from a low phosphorus diet was also reported in rat.[23] A recent study in rats fed on low phosphate diet showed that the calcium oxalate urolithiasis which developed was comparable with the bladder-stone disease.[24] Therefore, low dietary phosphate intake should play an important role in bladder-stone formation in north-east Thailand.

As a preventive measure

The beneficial effects of orthophosphate on the mineralizing capacity of urine tested in rachitic rat cartilage *in vitro* was described by Howard *et al.* in 1962.[25] Fleisch and Bisaz[26,27] subsequently indicated that pyrophosphate inhibited both hydroxyapatite and calcium oxalate precipitations. Since urine prophosphate can be increased by oral orthophosphate supplementation,[19] it should play an important role in the prevention of urolithiasis.

Long-term clinical trials with oral doses of orthophosphate in adult patients with metabolically active urolithiasis due to idiopathic renal lithiasis have been reported.[28] Orthophosphate, most often used as a neutral salt, was given orally in three of four divided doses to provide 1.5–2.0 g of phosphorus/24 hours. In over 90 per cent of the patients, complete control of stone formation was established. The effectiveness of therapy was the same in patients with or without hypercalciuria.

It is quite fascinating from our studies that orthophosphate supplementation can practically eliminate oxalcrystalluria and crystal clumping despite continuing ingestion of the oxalate-containing vegetables.[15] Therefore, the administration of oral orthophosphate for the prevention of bladder-stone disease has been carried out in north-east Thailand.

Hypothesis of stone formation in Thailand

The epidemiological, nutritional, biochemical, and clinical data during the period of study from 1963 to 1974 revealed the following pertinent findings:
1 Early introduction of rice supplements and low breast milk intake lead to low dietary phosphate intake, resulting in low urinary phosphate excretion. The renal tubules will excrete more ammonia to compensate for the low buffering of hydrogen ions from the low excretion of phosphate.
2 Consumption vegetables with a high oxalate content induces hyperoxaluria. Some degree of dehydration existing in the village infants and young children will aggravate the supersaturation of calcium oxalate, which finally leads to homogeneous nucleation and crystallization.
3 High urinary uric acid excretion was also found which is likely to reduce the inhibitory activity of the acid mucopolysaccharides (AMPS)—inhibitors of calcium oxalate crystallization.[29] Low pyrophosphate content in the urine will also reduce the protection against crystal growth and aggregation.

The possible mechanism of stone formation in Thailand is shown in Fig. 7.2. Primary homogeneous calcium oxalate nucleation and crystallization are initiated from supersaturation of calcium oxalate. Ammonium acid urate, which results from the combination of urinary uric acid and ammonia, may bind with calcium oxalate to form secondary heterogeneous nucleation and crystallization resulting in heavy crystalluria and clumping. The low inhibitory activities of AMPS and pyrophosphate will further enhance the mixed crystals to aggregate and become bigger. If this large and aggregate particle remains in the bladder for a period of time, it will become critical and may form a nucleus of stones.

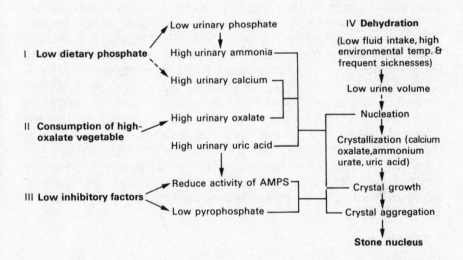

FIG. 7.2 Hypothesis of bladder-stone formation in Thailand.

This mechanism is supported by previous epidemiological data and urinary studies which showed very high incidence of oxalate crystalluria among village children; however, the crystalluria was usually in a mild to moderate degree without clumping. If the crystalluria became heterogeneous—comprising calcium oxalate, ammonium acid urate, and/or uric acid—the crystals would aggregate and become big; hence, the village children might complain of the presumptive stone symptoms of sandy and cloudy urine, dysuria, and occasional haematuria. The prevalence rate of presumptive symptoms was 2.6 per cent. In addition, 1.2 per cent either spontaneously passed small stones or required surgical removal, which might indicate that those particles which had remained in the bladder had aggregated into a nidus of stones.

There are several factors in the process of stone formation, which include variations in dietary habits and intakes, state of hydration, anatomy of urethra (male or female), and other unknown factors. This may explain the differences in the severity of stone symptoms and in the natural course of bladder-stone disease.

Preventive programme

From the proposed hypothesis indicates that low dietary phosphate plays a very important role in stone formation. This is also supported by previous animal experiments as well as the beneficial effect of orthophosphate in humans in the prevention or renal stone recurrence, as previously described. A field preventive programme was implemented in 1974 to test the effectiveness of oral orthophosphate supplementation in: (a) protection against crystalluria; (b) prevention of the occurrence of presumptive and positive stone symptoms; and (c) reduction in the incidence of bladder-stone disease.

The study covered approximately 1200 infants and pre-school children living in nine villages of Ubol province. Oral orthophosphate supplementation of 60 mg phosphorus/kg body weight per day was given daily under close supervision for a period of 5 years. If the results of this field trial prove to be useful, oral orthophosphate could be used to prevent bladder-stone disease in Thailand and perhaps in other countries where bladder stone is endemic and low phosphate intake is an important aetiological factor.

For long-term preventive measures, nutrition education and upgrade of nutritional status of pregnant and lactating mothers, infants, and pre-school children will reduce the prevalence and eventually eradicate this disease.

Concluding remarks

Idiopathic bladder-stone disease, one of the oldest recorded diseases of mankind, used to be common in young children of Great Britain, Europe, and north America before the nineteenth century. It continues to be a major medical problem in the middle-east, far-east, and south-east Asia. In Thailand, the disease occurs most commonly in small children under 5 years of age living in rural areas of the north and north-east.

Studies in Thailand during the past 15 years—including epidemiological, nutritional (dietary habits and intakes), life style, biochemical, and clinical aspects—have revealed that dietary factors may initiate bladder-stone formation. The findings indicate that low dietary phosphate intake and high oxalate consumption are crucial for the initiation of stone formation. The mechanism by which bladder stone is formed in Thailand has been discussed. Oral orthophosphate supplementation of the usual amount of intake (60 mg phosphorus/kg body weight per day) completely eliminates oxalcrystalluria and crystal clumping. Crystallization, crystal growth, and crystal aggregation are likely to be an early phase which precedes stone formation. Oral orthophosphate supplementation should thus be effective in the prevention of bladder-stone disease.

A field preventive programme was implemented in 1974 covering approximately 1200 infants and pre-school children living in the endemic stone area of Thailand. Oral orthophosphate supplementation proves to be effective in the prevention of crystalluria as well as bladder-stone disease, phosphate supplementation may be useful in preventing this disease in Thailand and other parts of the world.

Acknowledgements

Supported in part by the National Institute of Health, US Departments of Health, Education and Welfare, Grant No. AM 13066-01-8.

References

1. Smith, L. H., Jr. Symposium on stones. *Am. J. Med.,* **45**, 649 (1968).
2. Halstead, S. B. In *Idiopathic urinary bladder stone disease* (ed. R. Van Reen). Fogarty International Center proceedings, No. 37. Fogarty International Centre, Washington, DC, p. 121.
3. Gershoff, S. N., Prien, E. L., and Chandrapanond, A. Urinary stones in Thailand. *J. Urol.,* **90**, 285 (1963).
4. Van Reen, R. (ed.). *Proceedings of the WHO regional symposium on vesical calculus disease, Bangkok, Thailand.* VS Departments of Health, Education, and Welfare, Baltimore, Md., DHEW Publication No. (NIH) 77–1191 (1972).
5. Unakul, S. Urinary stones in Thailand, a statistical survey. *Siriraj Hosp. Gaz.* **13**, 199 (1961).
6. Chutikorn, C., Valyasevi, A., and Halstead, S. B. Studies of bladder stone disease in Thailand. II. Hospital experience. *Am. J. clin. Nutr.* **20**, 1320 (1967).
7. Halstead, S. B. and Valyasevi, A. Studies of bladder stone disease in Thailand. III. Epidemiologic studies in Ubol province. *Am. J. clin. Nutr.* **20**, 1329 (1967).
8. Valyasevi, A., Halstead, S. B., Pantuwatana, S., and Tankayul, C. Studies of bladder stone disease in Thailand. IV. Dietary habits, nutritional intake and infant feeding practices among residents of a hypo- and hyper-endemic area. *Am. J. clin. Nutr.* **20**, 1340 (1967).
9. Valyasevi, A., Halstead, S. B., and Dhanamitta, S. Studies of bladder stone disease in Thailand, VI. Urinary studies in children, 2–10 years old, resident in a hypo- and hyper-endemic area. *Am. J. clin. Nutr.* **20**, 1362 (1967).
10. Valyasevi, A. and Dhanamitta, S. Studies of bladder stone disease in Thailand. VII. Urinary studies in newborn and infants of hypo- and hyper-endemic area. *Am. J. clin. Nutr.* **20**, 1369 (1967).
11. Valyasevi, A. Protein-energy malnutrition in Asia with special reference to Thailand. In *Proceedings of the Third Asian Congress of Pediatrics, Bangkok, Thailand.* Bangkok Medical Publisher, pp. 66–72.
12. Halstead, S. B., Valyasevi, A., and Umpaivit, P. Studies of bladder stone disease in Thailand. V. Dietary habits and disease prevalence. *Am. J. clin. Nutr.* **20**, 1352 (1967).
13. Halstead, S. B. Cause of bladder stone in olde England—A retrospective epidemiological study. In *Proceedings of the Fourth International Symposium on Urolithiasis Research*, Williamsbury, Va., USA, June, 1980, Plenum Press, New York pp. 325–8.

14. Dhanamitta, S., Valyasevi, A., and Susilavorn, B. In *Idiopathic urinary bladder stone disease* (ed. R. Van Reen). Fogarty International Centre proceedings, No. 37. Fogarty International Centre, Washington, DC, p. 151 (1977).

15. Valyasevi, A. and Dhanamitta, S. Studies of bladder stone disease in Thailand. XVII. Effect of exogenous source of oxalate on crystalluria. *Am. J. Clin. Nutr.* **27**, 877 (1974).

16. Van Reen, R. Unpublished data (A. Loutfi and A. A. Rahman) (1977).

17. Valyasevi, A. and Dhanamitta, S. Diet and bladder stone disease. In *Proceedings of the Eleventh Congress of Nutrition*, Rio de Janero, Brazil, August, 1978 (In press).

18. Dhanamitta, S., Valyasevi, A., and Van Reen, R. Studies of bladder stone disease in Thailand. IX. Effect of orthophosphate and fat-free powdered milk supplementation on the occurrence of crystalluria. *Am. J. clin. Nutr.* **20**, 1387 (1967).

19. Valyasevi, A., Dhanamitta, S., and Van Reen, R. Studies of bladder stone disease in Thailand. X. Effect of orthophosphate and non-fat dry milk supplementation on urine composition. *Am. J. clin. Nutr.* **22**, 218 (1969).

20. Schneider, H. and Steenbock, H. Calcium citrate uroliths on a low phosphorus diet. *J. Urol.* **43**, 339 (1940).

21. Morris, P. G. and Steenbock, H. Citrate lithiasis in the rat. *Am. J. Physiol.* **167**, 698 (1951).

22. Sarger, R. H. and Spargo, B. The effect of low phosphorus ration on calcium metabolism in the rat with the production of calcium citrate urinary calculi. *Metab. clin. Exp.* **4**, 519 (1955).

23. Coburn, S. O. and Packett, L. V., Jr. Calcium, phosphorus and citrate interactions in oxalate urolithiasis produced with a low phosphorus diet in rats. *J. Nutr.* **76**, 385 (1962).

24. Warness, P. G., Knox, F. G., and Smith, L. H. Low phosphate diet in rats: A model for calcium oxalate urolithiasis. In *Proceedings of the fourth international symposium on urolithiasis research*. Williamburg, Va., USA, June 22–26, 1980, Plenum Press, New York, pp. 731–4.

25. Howard, J. E., Thomas, W. C., Jr., Mukai, T., Johnston, R. A., Jr. and Pascre, B. J. The calcification of cartilage by urine and a suggestion for therapy in patients with certain kinds of calculi. *Trans. Assoc. Am. Physicians* **75**, 301 (1962).

26. Fleisch, H. and Bisaz, S. Effect of orthophosphate on urinary pyrophosphate excretion and the prevention of urolithiasis. *Lancet* **1**, 1065 (1964).

27. Fleisch, H. and Bisaz, S. The inhibitory effect of pyrophosphate on calcium oxalate precipitation and its relation to urolithiasis. *Experimention* **20**, 276 (1964).

28. Smith, L. H., Thomas, W. C., Jr. and Arnand, C. D. Orthophosphate treatment in calcium renal lithiasis. In *Urinary calculi; recent advance in etiology, renal structure and treatment* (eds. L. C. Delatte, A. Rapado, and A. Hodgkinson). S. Karger, Basel, p. 188 (1973).

29. Robertson, W. G. Saturation–inhibition index as a measure of the risk of calcium oxalate stone formation. In *Idiopathic urinary bladder stone disease* (ed. R. Van Reen). Fogarty International Center Proceedings. No. 37. Fogarty International Center, Washington, DC, pp. 55–71 (1977).

8   Transnational adoptions

YNGVE HOFVANDER and
CLAES SUNDELIN

Introduction

The phenomenon of adoption is age old. In the extended family system a child who lost its mother has always been taken care of and raised by other relatives. In some cultures a child who has been breast fed by another mother than its own may be considered as if blood related.

Legal adoption within a certain country is also of old age, at least in most western countries. The procedures and restrictions vary considerably, as does the number of adoptions.

In several European countries the number of children available for adoption has decreased markedly during the last couple of decades. A number of factors are responsible for this development. There has been an unprecedented socioeconomic development which is the basis for an improved ability to care for one's own children. There are few destitute families or mothers these days compared with previous times. A social network has been developed to care for the few who for various reasons may be unable to care for themselves and their offspring. Another important factor is most probably the liberal abortion laws in many countries enabling more or less only 'wanted' children to be born. If a pregnancy occurs at the wrong age, at the wrong time, or from the wrong man it may be terminated by induced abortion. In Sweden at the present time every fourth pregnancy is aborted. Parallel to this development the number of children available and eligible for adoption has decreased sharply.

Transnational adoption is a fairly recent phenomenon. Although there is a certain exchange between industrialized countries the main stream of children is from developing to developed countries. One of the initiating events seems to have been the Second World War but in particular the Korean war, which left a large number of destitute mothers and abandoned children. Other recent wars or political instability such as in Vietnam or Kampuchea increase these numbers markedly.

As a result of these events but also of the fact that a large 'demand' of children for adoption in 'the West' has become apparent, numerous small orphanages have been created in many developing countries to take care of abandoned children for later adoption. Some of these are run by national or international charity organizations but others by a lawyer, a nurse, a social worker, or by a doctor who may have specialized in this field. The general impression by adoptive parents who come to take their child from these places is with few exceptions favourable. The financial support from adoptive parents and adoption societies may help to keep the standard at an acceptable level. There are exceptions though—large, overcrowded institutions with little possibility of giving each child a necessary personal attendance.

Number of transnational adoptions

Only few countries have a national registration of adoptions. Table 8.1 gives some idea about the yearly number of transnational adoptions in selected countries.[1]

TABLE 8.1   *Approximate yearly number of transnational adoptions in selected countries*

Sweden	2000
Denmark	1000
Holland	1000
Belgium	500
Norway	300
Total	4800

It is difficult to extrapolate with certainty as to the total number of transnationally adopted children per year. The listed countries may be particularly 'active'. However, a guess may arrive at figures in the order of 10 000–20 000/year. In view of this it is indeed remarkable that only a few studies have been published on the health and adaptation of these children.[2-4]

The motive for adoption

The most common reason for an adoption is involuntary childlessness—60–70 per cent in different studies.[5,6] About one-fifth state that they 'want to make an aid contribution' or that 'it is irresponsible to deliver more children of your own in an already overpopulated world'.[5] Another one-fifth are scared of another pregnancy, of the risk of malformations, or simply would like a fresh start again.

Even in carefully conducted interviews, however, it may be difficult to catch the *real* motive. Often it is not even clear or spelled out to the parents themselves. There may be mixed reasons or the husband and wife may have different motives.

Preparation for adoption

Adopting a child is not a simple matter. First of all the husband and wife have to agree among themselves. If childlessness is the reason, usually several years of frustrating investigations (and waiting!) have preceded the final decision, which for most people is not an easy one.

Most countries require a social investigation to ensure that the parents are 'suitable'—in itself experienced as a humiliating process, particularly as it is not required for biological parents to be. During the investigation the real reasons for the planned adoption will be analysed (including by the parents themselves!). Occasionally it becomes obvious that the situation calls for other remedies than a child and the social investigation has rather come to serve as a psychotherapy.

A large number of legal documents will have to be prepared, translated into English, and sent to the respective authorities in the donating country—all requiring a considerable administrative experience and power of endurance on the part of the parents. A positive selection is therefore inevitable.

The formal requirements to be accepted as adoptive parents in the recipient countries vary. In general parents should be of child-bearing age. Donating countries usually have much more specific requirements, which vary from country to country—the parents' age, religion, number of years of marriage (few countries will accept single parents), childlessness, previous children, and so forth. For instance, some countries may require that the parents have no previous children, while others require the opposite.

The first few months

The initial health screening should be aimed at:

1 Diagnosing malnutrition and infectious diseases which may be harmful to the child or its environment, e.g. salmonella.
2 Detecting hidden or obvious handicaps, disabilities, and delayed psycho-motor development.
3 Giving a detailed information to the parents on the child's physical and mental condition and how it may be expected to change.

Adoptive parents are usually much more anxious about the child's condition and future development than biological parents and may need support for a long time. As most adopted children are coming from developing countries the pattern of health on arrival is naturally similar to that in the country of origin. Table 8.2 shows some major findings in 105 children below 5 years from twenty developing countries.[5]

Most of the ailments are such that they can easily be dealt with and treated successfully in a month or two. Occasionally, however, some children continue to be salmonella carriers for months or years, which may put a great strain on the family as the child may be required—as in Sweden—to be isolated from other families and children.

TABLE 8.2 *Major findings on initial health screening in 105 adopted children in Sweden from twenty developing countries*[5,7]

Condition	Children affected (%)
Malnutrition (PEM) (slight–severe)	29
Weight <80 per cent of standard[10]	41
Delayed psychomotor development	18
Diarrhoea	20
Respiratory infection	18
Skin infection	18
Scabies	14
Giardia lamblia	14
Salmonella	12
Ascaris	9
Trichuris	8

Adaptation to family and society

What may be of greater interest than the physical health is how the children adapt—during the first few weeks, and later. In an extensive in-depth follow-up study which included a lengthy taped interview with the parents and an examination of the child's development some interesting findings were made. Table 8.3.[5]

TABLE 8.3 *Behaviour problems on arrival and at follow up 1–5 years later (n = 144)*

Problem	On arrival (%)	Later (%)
Disturbed sleep	49	11
Feeding problems	52	10
Defiance, aggressiveness	12	16
Display of temperament	16	20
Separation anxiety	22	14
Apathy	16	1

In the first few weeks the main problems concerned feeding and sleep. About 50 per cent of the children had disturbed sleep. Quite a few woke up repeatedly or cried hysterically, and refused to sleep in their own bed but insisted on sleeping with the parents. Some filled their bed with their belongings as a consolation. For the majority these problems disappeared after a couple of months.

About half the children were noted to have 'feeding problems', for the most part consisting of an enormous hunger which did not seem possible to satisfy. 'He was eating as if each meal was the last' as one mother expressed it. Only very few reacted by refusing to eat. The ravenous appetite seemed to coincide with a rapid catch-up growth and decreased after a few months.

Fig. 8.1 shows an example of the remarkable catch-up growth in weight and height which occurs in nearly all of these children regardless of age and not only in obviously malnourished or grossly underweight children. During this phase infants may consume 150–200 kcal/kg[7] as against 100–120 kcal/kg normally. Once the rapid growth spurt ceases the appetite returns to normal. The dramatic growth spurt seems to be caused by a combination of several factors, among them absence of infection, nutritious food, and—what may be most important—a lot of love and stimulation.[4,7]

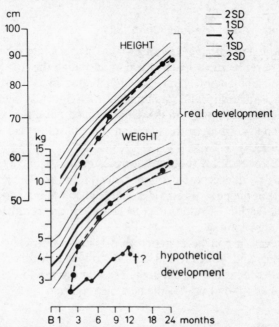

Fig. 8.1. Catch-up in a malnourished Indian boy adopted in Sweden at 2 months of age. At 2 years physical and psychomotor development was normal.

Defiance, aggressiveness, and display of temperament were common phenomena on arrival and they tended to remain a problem. The children who displayed apathy on arrival, however, quickly improved.

As a summary it may be said that about one-third of the children just 'walked in' to their family and adjusted momentarily as if they had always been part of it, while two-thirds did have adjustment problems which tended to be of a more prolonged nature, particularly for the children who were beyond infancy on arrival. However, when compared with a cohort of Swedish children of similar age attending child health centres there was very little difference in terms of behaviour problems. Similar conclusions can be drawn from a Danish study of 168 foreign-adopted children.[8] After 2 years in the country the children were well adapted and there were no more school problems than for Danish children.

The adaptation during school age may be of particular interest. In an extensive in-depth study of 207 children aged 10–18 years who had been in Sweden for many years it was found that these children in general performed as well as Swedish children. They were not bullied more than others. Interestingly, children who had arrived at the age of 1½–4 years did have more language difficulties than those who were younger *or* older on arrival, indicating that this age span may be a particularly sensitive period for erasing one language and building up a new one.[9]

The future of transnational adoptions

Adoption should have one single aim: it is a child who needs parents. It should never be the other way round, although the two aims may coincide.

For various obvious reasons national adoption should be aimed at in the first place and transnational should be a last resort. Present experience shows that thousands more families in developed countries would be able and willing to take children for adoption, were it not for various legal and other restrictions.

There are certainly thousands of children for whom a transnational adoption may be an acceptable (but not ideal) solution. The generally favourable outcome in the new country is certainly encouraging and should be brought to the attention of authorities in charge in the countries from which the children are coming.

However, it can be assumed that many countries with increasing self-sufficiency and wealth will more and more be able to arrange for the care of abandoned or destitute children. For some years to come, though, transnational adoption may be 'the next best solution'.

References

1. Ministry of Social Affairs, Sweden. *International adoptions*, vol. 6. Ministry of Social Affairs, Stockholm (1978).
2. Simon, R. J. and Altstein, H. *Transracial adoption*. Wiley, New York (1977).
3. Tizard, B. *Adoption: a second chance*. Open Books London (1977).
4. Winick, M., Meyer, K. K., and Harris, R. C. Malnutrition and environmental enrichment by early adoption. *Science* **190**, 1173 (1975).
5. Gunnarby, A., Hofvander, Y., Sundelin, C., and Sjölin, S. The health and adaptation of Non-European adopted children in Sweden. [In Swedish.] *Läkartidningen* (In press)
6. Hoksbergen, R. A. C. (ed.). *Adoption of children from distant countries*. Deventer (1979).
7. Hofvander, Y. Unpublished observations.
8. Pruzan, V. *Born abroad—adopted in Denmark* (summary in English). Social research institute, Copenhagen, No. 77 (1977).
9. Gardell, I. *Internationella adoptioner* (International adoptions). Allmänna Barnhuset, Stockholm (1980).
10. Jelliffe, D. B. *The assessment of the nutritional status of the community*. WHO, Geneva (1966).

9 Epidemic hysteria in schoolchildren: a world-wide syndrome

ALAN NORTON

On Wednesday November 8th 1978, sixty-two pupils attending a primary school in Kingston, Jamaica, became ill. The commonest symptoms were abdominal pain and of feeling unwell. A few children had vomited or had diarrhoea, some of the more severely affected had fainted, and some had been seen to be overbreathing. Sixteen children were taken to hospital, but only five were admitted and none was kept overnight. The next day there were thirteen new and similar cases, and twenty others suffered a recurrence of symptoms. While investigations into the cause of the outbreak were pursued the authorities decided to close the school on Friday November 10th. No cases were reported from the children's homes over the weekend, but when the school reopened on Monday 13th there were sixteen new cases and twenty-nine recurrences on that day. However, the next day there were only five cases, all new, and in the following week or so the outbreak, which started in October, gradually petered out.

Of the 890 girls in the school 159 were affected (18 per cent), whereas only thirty-seven boys were ill out of a total of 773 (5 per cent). Older children aged 10–11 were the most often affected and the youngest least. Members of the teaching staff recalled that sporadic cases of a similar type had occurred in the two weeks before the alarming outbreak on November 8th. A very full epidemiological investigation was mounted; contamination of food and water, chemical toxins, gases, and common infections were effectively ruled out as possible causes. One teacher noted that the children with recurrences were among the most troublesome pupils. Although there was no evidence at all of widespread anxiety or of distressing and upsetting recent events, several of the children did mention that they had been troubled by the dead school watchman's ghost or 'duppy'. However, he had died over a year previously.

This outbreak, which happened quite recently and has been comprehensively reported,[1] can usefully serve as a prime example of epidemic hysteria in schoolchildren. The report mentions almost all the features common in greater or lesser degree to so many of the outbreaks—not that such epidemics are very numerous—reported from all over the world in the last few decades. The cultural setting does, or course, strongly influence the details of the clinical picture, but in whatever continent the school is situated published reports stress at least some elements to be found in the list that follows:

Onset—This may be explosive or slowly cumulative or, as in the Jamaican episode, the outbreak though seemingly explosive turns out in retrospect to have had slow beginnings. The epidemic may spread from a herald case who is physically or mentally ill. For example, in an epidemic in Uganda affecting forty pupils and reported in 1973,[2] the herald case was suffering from schizophrenia. In the Jamaican outbreak, too, some of the early cases were considered in retrospect to have suffered from gastroenteritis. The type of onset of epidemics of hysteria in schoolchildren and others in the last 100 years has been classified minutely in Sirois's study[3] of the world medical literature.

Sex distribution—In the great majority of recorded epidemics far more girls than boys have been affected, the ratio often being as high as 4:1 in mixed schools.

Age distribution—The commonest age group affected by such illnesses in secondary schools is 12–14 years. In primary schools the heaviest effect is on the top form of 10–11-year-olds. Later in the epidemic the illness spreads to younger children.

Staff are seldom affected, rarely more than one or two in a school compared with scores of girls. But in one epidemic in Malaysia, in which seventy-eight pupils suffered, five teachers became ill.[4]

Parents and siblings at home—In most of the reported epidemics parents and younger siblings have rarely been affected. Occasionally, however, the epidemic has been spread through contact to another school, for example, in a series of epidemics in 1962 in Tanzania.[5]

Symptomatology—Abdominal pain, headache, nausea, faintness, feelings of dizziness, and weakness are the common symptoms in these epidemics. Others—diarrhoea, vomiting, shivering, and tremor—are less common. Rare prominent symptoms have included itching in a school in the southern USA in 1973,[6,7] coughing and sneezing in 1968 in a school in the eastern USA,[8] and convulsions in epidemics both in France[9] and Canada.[10] If, as quite often happens, overbreathing is a part of the epidemic picture, then the children experience its physiological consequences—numbness, tingling, and tetany. In Asia and Africa more bizarre symptoms and behaviour are experienced—screaming, weeping, and wailing in Malaysia;[11] wailing and giggling in Ghana;[12] laughing and running and being commanded by dead spirits in Uganda and Tanzania.[13]

Numbers—The number of children affected in these epidemics can be very large. In the two well-known epidemics fully described by McEvedy and others in English schools in the 1960s,[14,15] 180 out of 550 and 104 out of a total of 272 girls, respectively, were affected. But Sirois, who in an earlier paper[3] only considered outbreaks affecting twenty-five or more children, states[16] that of the smaller epidemics most escape notice. They are tacitly ascribed to gastroenteritis or to an unspecified virus.

Relapses and repeated attacks—One of the most characteristic features of these epidemics is the almost total disappearance of fresh cases of the illness at home at the weekend. On Monday it reappears, and as time goes on fewer fresh cases are reported and a higher proportion of the total are relapses. This pattern has been so consistently described in epidemics of hysteria that one is tempted to regard it as a hallmark and to question the authors' conclusions about an outbreak in three schools in Coventry, England,[17] in 1964. The outbreak affected 404 pupils and two staff members and was attributed to an 'infectious (probably viral) origin'. But the authors admit that in a small number of cases the symptoms were hysterical, and they do not explain the almost total absence of cases at the weekend or the high relapse rate.

Negative results of investigations—In individual pupils symptoms greatly outweigh physical signs and such physical signs as are shown are the outcome either of anxiety or of overbreathing. Investigations necessary to exclude bacterial infection of water and food, viral infections, toxic gases from heating systems, and poisoning by insecticides must of course be made, but some of these laboratory investigations may take a long time.

Staff alarm, parental anxiety, media attention, police, firemen, inspectors, etc.—Any sizeable outbreak is bound to have some at least of these public consequences, which enhance anxiety and add to the drama of the event. The number of cases in consequence rises steeply. Nitzkin[18] gives a vivid description of the effect of the arrival of firemen to investigate the possibility of 'poison gas' accounting for one 11-year-old girl's sudden illness in a school in Florida. In his report on the epidemic in Ghana already referred to, Adomakoh[12] specifically states that the attention of the media to the wailing and uncontrolled laughter exacerbated the epidemic. Agitated parents contributed, it seems, to the spread of an epidemic of twitching and jerking in a Louisiana school in 1939.[19]

Stress—In many, perhaps in most, of the epidemics reported the children have been under a strain. This strain can range from the jejune, the imminence, say, of examinations, to the severe. In the Blackburn epidemic already alluded to[14] the town had already experienced a poliomyelitis epidemic earlier in the year; there were reports of lorry drivers refusing to deliver to the town and of landladies elsewhere refusing bookings for holidays. The day before the outbreak many girls had attended the cathedral and had had to wait outside for 3 hours because the personage who was to open the ceremony was late in arriving. During this wait twenty girls either fainted or had to lie down on the ground. Less specific and less convincing kinds of stress are referred to by a number of other authors: girls in a Ghana school[12] dissatisfied with the food and living conditions; rapid social change in Malaysia[20] and a very troublesome headmaster; rapid socioeconomic change in East Africa[13] and conflict between forward-looking policy at school and traditional culture at home; an authority figure in another Malaysian school accused of favouritism;[11] pressure to achieve academic success and great disparity in age within forms at a school in Uganda;[2] tug-of-war in Uganda, again, between home and school.[21]

Other points—In many outbreaks simple measures such as getting the children to lie down are enough. However, many children have to be sent home and some may have to be sent to hospital and perhaps admitted for observation and investigation, particularly if the nature of the outbreak is obscure. Rapid recovery may be the general rule; but, especially if over-breathing has been observed, symptoms and signs may persist for several days. In the Blackburn epidemic,[14] for example, by day four of the outbreak twenty of the girls were still in hospital, were continuing to faint or over-breathe to the point of tetany. In another English epidemic[22] a quarter of the affected children still had some symptoms after 72 hours.

The illness may sometimes be seen to spread by direct vision. Polk,[7] for example, reporting an epidemic of which itching was the most prominent symptom says: 'Students became ill after seeing other students manifesting signs of the reported illness' and Levine *et al.*[6] commented on a visual chain of transmission. Figueroa, too, in the Jamaican epidemic[1] says that when children were being examined others were 'attracted by the activity and com-plained of symptoms although they appeared quite well'.

Several authors have stated, perhaps not surprisingly, that children known to show neurotic symptoms[8,14,17] and conduct disorder,[1,10,15] or known to be attending a child guidance clinic,[14,15] are those most likely to be caught up in an epidemic and in relapses. Others have noticed that those of low socio-economic status, the poor, are more likely to be affected. For example, Sirois[3] in his very extensive study of the literature of epidemics in schools and other institutions (factories, hospitals) concluded that such people are over-represented. But this is not a finding mentioned by the majority of authors.

Remedies

The clue to the most effective remedy lies in the observation, so repeatedly made in so many epidemics, that the number of cases falls off dramatically at weekends. The drastic step of closing the school is highly successful, but it does need courage, because the decision has to be made on a combination of rather fallible positive evidence of the hysterical nature of the outbreak together with negative evidence from results of laboratory and other investi-gations which may well be slow to come in. So long as doubt about the diag-nosis remains, tension mounts, rumour is rife, and anxiety spreads from the school to the parents and the community, and resonates because of the interest of the media. In the Jamaican outbreak[1] a food or water-borne illness and the effect of an environmental toxin were sufficiently excluded within 48 hours to allow a reasonably sure diagnosis to be made. The school was shut for 3 days, a period perhaps not quite long enough, for sixteen new cases and twenty-nine recurrences were experienced on the day of reopening. But the outbreak was effectively over. In the Blackburn epidemic[14] the school was shut for 6 days and again there was a brief recrudescence on its reopening, with sixty girls affected, but the outbreak was effectively over. Closing of the school was effective in both the Alabama epidemic[6,7] and in Ghana[12] and

Tanzania.[5,13] The events at the school in Malaysia,[11] which seemed to be started on several successive days by a particular girl, were not effectively dealt with by exorcism performed by a *bomoh* or witch-doctor. But closing the school for a few weeks did succeed. The outbreak in schools in Coventry[17] was brought to an end in two of the three schools by closing them, though the nature of the epidemic remained and remains uncertain.

Positive signs of the hysterical nature of an epidemic are potentially numerous and naturally increase the longer the outbreak goes on. Negative results from clinical examination and from laboratories accumulate too. Sooner or later and preferably sooner someone has to take a firm decision— even if he is not quite sure—and tell the girls, the school, the parents, and the media that the outbreak is hysterical and if necessary that the school should be closed for a time. He will have to take the slight risk of being wrong and perhaps of spreading an infectious disease in the neighbourhood. The one who has to take the decision could do worse than heed the advice of Johann Weyer[23] in the Rhineland in the sixteenth century, when asked what should be done with bewitched and demoniacal people in convents: 'It is necessary first of all that they be separated and that each of the girls be sent to her parents or relatives.'

Any discussion of epidemic hysteria in schoolchildren runs head on into two obstacles: in what sense is the word hysteria being used; and is there anything special about hysteria in children that does not also apply to other groups?

The word hysteria

I make no apology for using this word, although its use has given rise to argument for decades, even to the extent that the illness has been denied existence by Slater.[24] The phrase 'epidemic hysteria' has been hallowed by long use, everybody has a fair idea of what it signifies, and no one has suggested a better. 'Mass hysteria', 'mass anxiety', 'communicated hysteria', 'epidemic transient situational disturbance', 'mental contagion', 'crowd hysteria', 'epidemic emotional disturbance', 'hysteria 1975', 'hysterical contagion', and, most recently of all, 'collective stress syndrome'[25] are some of the alternatives that have been used without advantage. Lay descriptions— laughing disease, laughing mania, running mania, mystery illness, mystery gas—have been used in describing some epidemics in schools, but the more extreme titles—'phantom' anaesthetist of Matoon',[26] 'phantom slasher of Taipei'[27]—described epidemics in adults.

The symptoms in the school epidemics constitute a complex and varied assortment. They may be organic or even psychotic in the early case or cases. The majority of children show anxiety symptoms perhaps amounting to panic, there may be overbreathing and its consequences arising from fear, and some conversion symptoms and dramatic behaviour. Some children may be histrionic in their personalities and be much more attention seeking than others. Some, too, will be able to escape from stress by adopting the sick role, for them a form of gain. But most, one must suppose, have no psychopathology and the illness is not to be understood in terms of the individual alone. The explanation is to be found in the psychology of groups, too large a subject to be entered into here, but crucially relevant for all those—heads of schools and community physicians—who are at risk of having to cope and take decisions in this field. It is essential to remember that symptoms and spread very seldom happen at home and illnesses are cured if the group is broken up and the children are sent home.

Hysteria in other groups

Similar phenomena to those under discussion have occurred in groups of other children. There have been reports, two from the USA and one from England, of epidemic hysteria in school marching bands. Pfeiffer[28] published an account of a high-school band, who were playing music at a football match, marching in heavy woollen uniforms after travelling 56 miles by bus in Maine. There had been an outbreak of botulism much reported in the press in the previous week. Eight girls, horn players, developed abdominal pains and numb hands. Several showed overbreathing and carpopedal spasm. Their team, too, was being defeated. In 1972 in England, in a much larger outbreak,[22] 130 visiting children, mostly belonging to juvenile jazz bands, developed abdominal pain while attending a gala. Several fainted or fell to the ground. The summoning of police and ambulance services exacerbated the affair, and within two hours 168 visiting children and four adults (but no local children) were taken to hospital. The children were all very frightened and bewildered, but in the hospitals were separated from each other and soon recovered. The commonest age in the children (only four were boys) was 13. In 1973 in the USA[29] fifty-seven members of a marching band—comprising members of both sexes and including one adult—after playing instruments at a football game, developed headache, nausea, and weakness. Six girls fainted. At half-time when the illness began the team favoured to win was losing. Again, heavy uniforms were being worn and the worst affected were wind players. The percentage of girls and boys affected was 57 and 36. Those three episodes certainly belong in this study for they happened to schoolchildren in a group, though not when they were in school.

This is the place to mention an event that occurred in July 1980 in the English midlands. It appears to present many of the classical features of epidemic hysteria. About 300 out of 500 children aged 9–14, taking part in a jazz-band competition, rapidly became ill and collapsed. From the information given in the lay press, the features of the outbreak can be roughly pieced together and recounted systematically. The onset was explosive; as one report has it, 'some of the children were catching their friends as they fell and then were falling themselves, the scene was like a battlefield, sprawled bodies everywhere'. Some parents became ill. The symptoms in the outbreak included nausea, vomiting, stomach pains, burning eyes, a metallic taste, trembling and sore throat. Some children relapsed and two were readmitted to hospital; apart from these two readmissions, only seven children were kept in hospital overnight. Stress can only be regarded as minor: the excitement of competition; a short coach journey; standing to attention in uniform. Immediate investigations were negative: crop-spraying with insecticide had not been carried out locally; herbicides had not been used; there were no 'fumes'; and many children had brought their own food so food poisoning was most unlikely. Parents were very naturally alarmed, and police, ambulances, firemen, and the media were much in evidence.

The report of the environmental health department of the local authority states that 255 names were recorded at hospitals, but more than 400 were seen. Of those seen and interviewed by environmental health staff later, thirty-seven were male and 187 female. Various possible causes of the incident were ruled out: food, fumes from a fire, water, pesticide spray, VHF radio waves, ultrasound, infrasound, and hypochlorite from toilets. Humid weather, standing for long periods before inspection, tight-fitting uniforms and hats, the report concludes, were precipitating causes of the outbreak.[30]

To confine the study to children of school age does, however, introduce boundaries that are somewhat artificial. Very similar phenomena have happened in fairly closed, mainly female, groups of slightly older people. These do merit a brief mention because they emphasize the width of Sirois's definition[3] and description and the links that the epidemics of this century have with those of earlier times:

The classical outbreak involves a small group of segregated young females; it appears, spreads and subsides rapidly, occasionally recurs, and is easily controlled with the dismemberment of the group. It is manifested by anxious and conversion reactions, sanctioned by the affected group.

In June 1962 in a textile plant in South Carolina,[31] fifty-nine females and three males believed they had been bitten by insects, but no one could see the creatures. Nausea, fainting, headache, dizziness, and cramps were the chief symptoms, and the illness originated in a 22-year-old girl who had fainted many times before. Eventually, all sufferers were sent home and the epidemic ended. In 1973 at a large TV assembly factory in Singapore with nearly a thousand employees,[32] eighty-three women and one man, all Malays, had between them ninety spells of illness. Symptoms included dizziness, numbness, faintness, 'hysterical' behaviour, screaming, beliefs about 'ghosts and djinns', and 'trances'. The job comprised monotonous, repetitive, simple operations. Ikeda in 1966 reported an epidemic of emotional disturbance in a leprosarium in Japan.[33] Eleven nurses were affected, the setting was one of extensive cultural change, and the morale of all the staff was low.

The epidemic that occurred in Indian children in northern Ontario in 1972[10] was unusual in a different way in that it came on at home, not when the children were at school. Children affected, who numbered more than thirteen, were of both sexes and ranged in age from 11 to 18. The stress was exceptional: cold, dark winters and a strict fundamentalist religious background among the parents. The children, on the other hand, had been exposed through school, radio, newspapers, and magazines to the culture of the south of the province. An appetite, created for unobtainable material objects, had been increased in the many children who had had a spell of boarding school in the south. A generation gap yawned. The children showed irritability and had seizures and intractable headaches. Siblings were commonly affected. A similar episode had happened 8 years before.

To go farther afield and back into history would take us too far, though the description in Hecker's book, quoted by Schuler and Parenton,[19] of an epidemic in a convent of many nuns mewing like cats sounds germane. From the same source comes a description in fifteenth-century Germany of nun-biting mania which eventually spread to Holland and even to Rome. Nor should we travel too far into transcultural psychiatry to consider those well-known culture-bound syndromes of Asia—amok, latah and koro—all of which can occur in epidemic form.

Incidence

How common are epidemics of hysteria in schools and schoolchildren? Unless numerous outbreaks go unpublished, which seems unlikely because of their clinical interest and the accompanying brouhaha, and unless many small outbreaks go unnoticed and have other labels attached, the answer is clear. They are most unusual. A fairly thorough search of the English language literature has turned up the list shown in Table 9.1.

In the western world at least, therefore, such epidemics are uncommon. I could find only these eighteen western school epidemics in the English language literature of the last 40 years. In his comprehensive monograph,[3] Sirois says that in the last quarter of the nineteenth century reports of epidemics of hysteria (not only, of course, in schools) were quite numerous, that from 1900 to 1940 there were few reports, but that since the war interest has revived. By 1974 he was able to gather only twenty-seven accounts in the last hundred years of epidemics affecting more than twenty-five people. This was from the world, not only the English language, literature. The epidemics occurred in young people in schools, factories, hospitals, and other institutions. Sirois specifically excluded organized contagious behaviour—ritual possession, Voodoo ceremonies, and performances like those of the whirling dervishes. Elsewhere he has stated his view that the smaller outbreaks may not only go unnoticed and unreported, but may also be not particularly uncommon. From the results of an enquiry in 1973 of all the schools of

TABLE 9.1 *Epidemics of hysteria in English references*

Place	Year of outbreak	Year of report	No. of females (%)	No. of males	Main symptoms	Trigger	Stress	Media present?	Reference
Coventry, UK, schools*	1964	1964	404 and 2 staff†		Headache, nausea, vertigo, 'collapse'			Yes	17
Blackburn, UK, school	1965	1966	About 180 (100)		Dizziness, faintness, overbreathing		Yes	Yes	14
Portsmouth, UK, school	1965	1966	About 108 new cases (100)		Vomiting, abdominal pain, faintness	A few with gastrointestinal infection			15
Belfast, UK, school	1969	1970	28 (>95)	1	Faintness, nausea, abdominal pain, headache	First girl, tonsillitis	Yes		34
Northumberland, UK, schoolbands	1972	1973	About 168 (>95)	4	Dizziness, abdominal pain, nausea, headache		Yes	Yes	22
London, UK, school	1972	1973	9 (100)		Faintness, falling	Yes			35
France, boarding school	1950	1952	5 (100)		Seizures				9
Louisiana, USA, school	1939	1943	7 (100)		Twitching, jerking	Yes	Yes		19
Louisiana, USA, school‡	1962	1965	21 (96)	1	Blackouts, over-breathing, faintness, tremor	Yes	Yes	Yes	36
Maine, USA, schoolband	1963	1964	8 (100)		Overbreathing, tetany, numbness, abdominal pain	Yes	Yes	Yes	28
Eastern, USA, school‡	1966–7?	1968	14 (87)	2	Crying, laughing, tremor, headache	Yes, girl killed in car accident			37
Eastern city, USA, school‡	1968	1972	47 (89)	6	Screaming, coughing, overbreathing, abdominal pain, burning throat				8

Place	Year of outbreak	Year of report	No. of females (%)	No. of males	Main symptoms	Trigger	Stress	Media present?	Reference
Alabama, USA, school	1973	1974	71 (72)	27	Itching, rash, headache, cough, nausea				6
Southern USA, schoolband	1973	1977	57 and 1 adult†	20	Headache, nausea, weakness, faintness		Yes		29
Florida, USA, school	1975?	1976	23 (68)	11	Headache, nausea, dizziness, pain	Yes	Yes	Yes	18
Quebec, Canada, school	1974	1975	11 (100)		Abdominal pain, nausea				16
Quebec, Canada, school	1974	1975	100 (100)		Headache, abdominal pain, nausea, over-breathing, laughing				16
Jamaica school	1978	1979	159 (81)	37	Abdominal pain, vomiting, faintness, overbreathing		Yes?		1
Ghana school	1967	1973	62 (100)		Weeping, wailing, laughing	Yes, herald case	Yes	Yes	12
Tanzania and Uganda schools; series of epidemics	1962 1971 1962	1968 1973 1963	95 (100) 0 164 (100)	40	Laughing, running, dead spirits. Laughing, self-neglect, grimacing, fear of ants	First patient schizophrenic	Yes		13, 21 2 5
Malaysia schools	1971?	1975 1963	78 (100) 8 (100)		Screaming, running, shouting, trance states, ghosts, fainting, weeping, wailing		Yes Yes	Yes Yes	4, 20 11

* The exact nature of this outbreak remains conjectural. Many features suggest hysteria, but the authors favour a basic virus infection with 'hysterical elements' present as well.
† Sex distribution not given.
‡ Non-white, USA school.

Quebec,[16] with the exception of Montreal, he calculated that there was about one outbreak per annum per thousand schools. This ratio, transferred crudely to the schools of England and Wales, would lead one to expect twenty-eight outbreaks each year. If they indeed happen on this scale, the public hears very little about them.

Other authors broadly agree. Schuler and Parenton, writing in 1943,[19] say that the 'medical literature [on epidemic hysteria] in the latter part of the nineteenth century is abundant', but they could not find a single publication in the USA for over 40 years. Levine *et al.*, writing in 1974,[6] could find only sixteen episodes recorded in the English-language journals since 1900. These concerned a total of 400 children and twenty-seven teachers.

On the other hand, in parts of Asia, Malaysia in particular, and east Africa, especially on the western shore of Lake Victoria, outbreaks seem to be quite common. Teoh[20] says that in the decade 1962–1971 twenty-nine schools in Malaysia had epidemics of hysteria and there were seventeen in 1971 alone. These episodes have been dramatically publicized by the press. In Tanzania in 1962–63 an episode[5] resonated between several villages for more than a year, affecting children as far as 50 miles away. The epidemic described by Adomakoh in Ghana[12] also dragged on and spread in the neighbourhood over a period of 3 years.

The consequences of outbreaks of epidemic hysteria being uncommon in western schools, or at least being seldom recognized as such, are important. Most teachers, including head teachers, even if they have read about such outbreaks may well not encounter one in the whole of their careers. The same is true of community physicians, the first doctors to be caught up in an epidemic. Psychiatrists, who could be expected from their training to know most about the subject (though they again will rarely have had first-hand experience), are most unlikely to be consulted early. Until a diagnosis has been made they have no reason to be.

Diagnosis

The symptoms complained of and the physical signs shown in epidemic hysteria, in common with those in individual hysteria, have become less dramatic over the years. In western countries dancing manias, tarantism, and mewing nuns are no longer encountered, just as co-conscious personalities and 'grande hystérie de Charcot' at or before the turn of the century and those gross paralyses and anaesthesias of the 1914–18 war have disappeared. Epidemics of disordered behaviour regarded by most people as within the sphere of psychiatry—laughing, running aimlessly, wailing, and screaming— are now largely confined to Africa and southern Asia; the diagnosis is evident and seldom gives rise to difficulty.

Elsewhere, as Merskey[38] has pointed out, schoolgirls and factory workers are familiar with the scientific notions of infectious illness, and some of the symptoms develop in conformity with these ideas. The symptoms suggest, superficially, infection by bacteria or viruses or poisoning, and no administrator or clinician in his senses is going to think first of hysteria; only when fairly simple investigations have proved negative are suspicions likely to be aroused and attention paid to features suggestive of epidemic hysteria.

Even so, many reports in the literature leave doubts unresolved. To arrive at a conclusive diagnosis is often not easy. Even after 25 years some slight doubts remain about the exact nature of the epidemic at the Royal Free Hospital in London in 1955. But opinion has swung behind an organic diagnosis[39] —'epidemic myalgic encephalomyelitis'—particularly after studies of similar and more recent outbreaks. Nevertheless it was curious that the epidemic was very largely confined to young female members of the staff[40] and that neither patients nor male staff caught it, and it may be significant that there was a frightening concurrent outbreak of poliomyelitis.

Several outbreaks have been reported in the literature of 'epidemic nausea and vomiting' in schools in England:

1 Reported in 1936, this outbreak affected a girls' boarding school[41] with 117 residents. There was sudden vomiting or nausea; one-quarter had giddiness but there was no fever. Half the girls and one-third of the adults were affected and the outbreak lasted less than 48 hours. There was no spread to the homes and no visitors were affected.
2 Reported in 1939, this affected a girls' day school with eighty-two girls and twenty staff.[42] There was dizziness, nausea, vomiting, and some cases with headache. A few had fever. Forty-five girls and four staff were affected, and there were a few cases in the surrounding countryside.
3 This outbreak, reported in 1943, affected a girls' boarding and day school and later a boys' school[43] with dizziness, nausea, and vomiting. There were some cases in the general population, and some cases relapsed.

These three epidemics seem ostensibly to be examples of a mild gastro-intestinal infection, but it is difficult to be sure.

The pattern of a sure 'organic' outbreak is exemplified in McEvedy's[15] description of an epidemic at a school at Wrexham, North Wales, in 1965. Infection arose from a member of the canteen staff, a symptomless carrier of Sonne dysentery organisms. At this mixed school large numbers of children failed to arrive at school rather than developing symptoms while they were there. In fact only 10 per cent of the ill children reported sick in school hours. The youngest pupils, 5-year-olds, were the first to show symptoms— abdominal cramp and vomiting—and the infection moved up the school age-ladder on succeeding days.

Another example of the organic type of outbreak is that reported by McMillan and Thompson,[44] which occurred at a boys' day school in south London in 1969. On the second day of the autumn term, seventy-eight boys became ill with malaise, aching limbs, diarrhoea, and vomiting. The symptoms came on after lunch but only after the boys had gone home. Hospital admission was needed for seventeen, twelve were comatose and collapsed, and three were dangerously ill. A very extensive epidemiological hunt excluded possible causes such as bacteria, lead, and organophosphorus compounds, and established that all those affected had eaten potatoes from a bag left over from the summer term. It was virtually certain that the epidemic was caused by the consumption of 'greened' potatoes. Solanine poisoning is possibly an unrecognized cause of other milder but mysterious epidemics of 'gastroenteritis'.[45]

But, in general, negative physical and laboratory findings, a high but not too high index of suspicion, and the presence of several of the features of epidemic hysteria outlined here should leave few doubts.

Theories

Most of the authors who have written the accounts of the epidemics summarized in this article have been content to describe the phenomena without concerning themselves very much with theoretical issues. They may have dwelt on stressful factors—physical, psychological, or cultural—and most will have noted that both the break-up of the group and frank speculation that the whole episode is hysterical bring the epidemic to a speedy end. Several authors have, however, gone much farther than this into the causes. Schuler and Parenton,[19] for example, in commenting on the epidemic they studied in Louisiana, remark that it occurred in the French-speaking Roman Catholic southern part of the State about the time of Mardi Gras and that festival's accompanying excitement. The community was more anxious than usual because of an epidemic of typhoid nearby. The first case had a need to take on the sick role because she was under severe stress. Strangely, although this original girl had high prestige, the next girls to be affected were not among her closest associates. Therefore, in this example—in contrast to some later epidemics—friendship networks did not predict the course of spread.

Other authors who have delved into the sociology of the epidemics they have described include Sirois,[3] Teoh,[4] Benaim *et al.,*[35] Armstrong and Patterson,[10] and Moss and McEvedy.[14] Kerckhoff *et al.*[47] studied the sociological features and the pattern of spread of the epidemic described by Champion *et al.*[31] Meerloo's[48] study of the clues and signals used in mental contagion goes much deeper and wider than the limits imposed by the title of this article. He notes the use of archaic forms of communication like dancing, and sees the universal compulsion to imitate as linking the hysterical outbreaks of the middle ages with Latah, the Jumpers and Shakers of nineteenth century Pennsylvania, and modern epidemics in schools in the west.

But it is to Gehlen's paper that those interested in theories of collective behaviour should turn.[46] Are the phenomena of epidemic hysteria best considered as an unpleasant consequence of stress or are they a type of craze? Are they an escape from stress into the sick role, and if so are they an escape for only a few or for many? Are early participants in an epidemic likely to be charismatic figures, and is this why the epidemic spreads down, not up, the age scale? How important are friendship networks in the spread of contagious behaviour? To these and other similar questions only the most tentative answers can as yet be given. With these speculations terra firma has been abandoned and scientific caution says stop.

Summary

A detailed description is given of an outbreak of hysteria in a school in Jamaica in 1978. This outbreak is typical in many ways. Typical features include: far more girls than boys affected; onset maximal in girls aged 12–14 years old; staff and people at home—parents and siblings—seldom affected; symptoms outweigh physical signs; relapses and repeated attacks become increasingly common; spread by sight of other children affected; overbreathing and its consequences may be observed; number affected may be large, e.g. 100–200; cases originate in school and subside at home or if the school is closed; presence of the media exacerbates the outbreak; and stress has preceded most outbreaks.

Outbreaks occur in Africa and south-east Asia, but they have special features. They may in certain areas and at certain times be endemic. Outbreaks reported in the English language journals in the west are uncommon. Smaller outbreaks may be much commoner and go unnoticed and unreported.

Brief details are tabulated of all the published epidemics of the last 20 years. Outbreaks in groups of young people other than schoolchildren are briefly mentioned, and diagnostic difficulties are outlined.

References

1. Figueroa, M. In *An epidemic of illness at a primary school in Kingston, Jamaica.* Caribbean Epidemiology Centre Surveillance Report, No. 4, pp. 1–5 (1979).
2. Muhangi, J. R. A preliminary report on 'Mass Hysteria' in an Ankole school. *E. Afr. med. J.* **50**, 304–10 (1973).
3. Sirois, F. Epidemic hysteria. *Acta psychiatr. scand.* Suppl., 252 (1974).
4. Teoh, J.-I., Seewondo S., and Sidharta, M. Epidemic hysteria in Malaysian schools: an illustrative episode. *Psychiatry* **38**, 258–68 (1975).
5. Rankin, A. M. and Philip P. J. An epidemic of laughing in the Bukoba district of Tanganyika. *Central Afr. J. Med.* **9**, 167–70 (1963).
6. Levine, R. J., Sexton, D. J., Romm, F. J., Wood, B. T., and Kaiser, J. Outbreak of psychosomatic illness at a rural elementary school. *Lancet* **2**, 1500–3 (1974).
7. Polk, L. D. Mass hysteria in an elementary school. *Clin. Pediatr. (Philadelphia)* **13**, 1013–4 (1974).
8. Goldberg, E. L. Crowd hysteria in a junior high school. *J. Sch. Hlth.* **43**, 362–6 (1972).

9. Michaux, L., Gallot H.-M., Lampérière, T., and Juredieu, C. Considérations psychopathologiques sur une épidemie d'hystérie convulsive dans un internat professionel. *Arch. Franç. Ped.* **9**, 987 (1952).

10. Armstrong, H. and Patterson, P. Seizures in Canadian Indian children: individual, family and community approaches. *Can. Psychiatr. Assoc. J.* **20**, 247-55 (1975).

11. Tan, E. S. Epidemic hysteria. *Med. J. Malaysia* **18**, 72-6 (1963).

12. Adomakoh, C. C. The pattern of epidemic hysteria in a girls' school in Ghana. *Ghana Med. J.* **12**, 407-11 (1973).

13. Ebrahim, G. J. Mass hysteria in schoolchildren. *Clin. Pediatr. (Philadelphia)* **7**, 437-8 (1968).

14. Moss, P. D. and McEvedy, C. P. An epidemic of overbreathing among schoolgirls. *Br. med. J.* **2**, 1295-300 (1966).

15. McEvedy, C. P., Griffith, A., and Hall, T. Two school epidemics. *Br. med. J.* **2**, 1300-2 (1966).

16. Sirois, F. À propos de la fréquence des epidémies d'hystérie. *Union Med. Can.* **104**, 121 (1975).

17. Pollock, G. T. and Clayton, T. M. 'Epidemic collapse': a mysterious outbreak in three Coventry schools. *Br. med. J.* **2**, 1625-7 (1964).

18. Nitzkin, J. L. Epidemic transient situational disturbance in an elementary school. *J. Fla. med. Assoc.* **63**, 357-9 (1976).

19. Schuler, E. A. and Parenton, V. J. A recent epidemic of hysteria in a Louisiana high school. *J. Soc. Psychol.* **17**, 221-35 (1943).

20. Teoh, J.-I. Epidemic hysteria and social change: outbreak in a lower secondary school in Malaysia. *Singapore med. J.* **16**, 301-6 (1975).

21. Kagwa, B. H. The problem of mass hysteria in East Africa. *E. Afr. Med. J.* **41**, 560-4 (1964).

22. Smith, H. C. T. and Eastham, E. J. Outbreak of abdominal pain. *Lancet* **2**, 956-8 (1973).

23. Weyer, J. In Zilboorg, G, (in collaboration with Henry, G. W.) *A history of medical psychology*. Norton, New York, p. 221 (1941).

24. Slater, E. Hysteria. *J. ment. Sci.* **107**, 359-81 (1961).

25. Mann, J. and Rosenblatt, W. Collective stress syndrome. *J. Am. med. Ass.* **242**, 27 (1979).

26. Johnson, D. M. The phantom anaesthetist of Mattoon. *J. Abnorm. Psychol.* **40**, 175-86 (1945).

27. Jacobs, N. The phantom slasher of Taipei: mass hysteria in a non-Western society. *Soc. Problems* **12**, 318-28 (1965).

28. Pfeiffer, P. H. Mass hysteria masquerading as food poisoning. *J. Maine Med. Assoc.* **55**, 27 (1964).

29. Levine, R. J. Epidemic faintness and syncope in a school marching band. *J. Am. med. Ass.* **238**, 2373-6 (1977).

30. Environmental Health and Control Department, Ashfield District Council. *The Hollinwell incident*. Ashfield District Council, Kirkby (1980).

31. Champion, F. P., Taylor, R., Joseph, P. R., and Heddon, J. C. Mass hysteria associated with insect bites. *J. S. Carolina med. Assoc.* **59**, 351-2 (1963).

32. Chew, P. K., Phoon, W. H., and Mae-Lim, H. A. Epidemic hysteria among some factory workers in Singapore. *Singapore med. J.* **17**, 10-15 (1976).

33. Ikeda, Y. An epidemic of emotional disturbance among leprosarium nurses in a setting of low morale and social change. *Psychiatry* **29**, 152-64 (1966).

34. Lyons, H. A. and Potter, P. E. Communicated hysteria. *J. Irish med. Assoc.* **63**, 377-9 (1970).

35. Benaim, S., Horder, J., and Anderson, J. Hysterical epidemic in a classroom. *Psychol. Med.* **3**, 366–73 (1973).
36. Knight, J. A., Friedman, T. I., and Sulianti, J. Epidemic hysteria: a field study. *Am. J. Publ. Hlth.* **55**, 858–65 (1965).
37. Helvie, C. O. An epidemic of hysteria in a high school. *J. Sch. Hlth.* **38**, 505–9 (1968).
38. Merskey, H. *The analysis of hysteria.* Baillière Tindall, London, p. 176 (1979). –
39. *British Medical Journal. Br. med. J.* **1**, 1437–8 (1978).
40. *British Medical Journal. Br. med. J.* **1**, 1–2 (1970).
41. Miller, R. and Raven, M. Epidemic nausea and vomiting. *Br. med. J.* **1**, 1242–4 (1936).
42. Gray, J. D. Epidemic nausea and vomiting. *Br. med. J.* **1**, 209–11 (1939).
43. Bradley, W. H. Epidemic nausea and vomiting. *Br. med. J.* **1**, 309–12 (1943).
44. McMillan, M. and Thompson, J. C. An outbreak of suspected solanine poisoning in schoolboys: examination of criteria of solanine poisoning. *Q.J. Med. New Series* **XLVIII**, 227–43 (1979).
45. *British Medical Journal, Br. med. J.* **2**, 1458–9 (1979).
46. Gehlen, F. L. Towards a revised theory of hysterical contagion. *J. Hlth. Soc. Behav.* **18**, 27–35 (1977).
47. Kerckhoff, A. C., Back, K. W., and Miller, N. Sociometric patterns in hysterical contagion. *Sociometry* **28**, 2–15 (1965).
48. Meerloo, J. A. M. Mental contagion. *Am. J. Psychother.* **13**, 66–82 (1959).

10 Nutrition education in a changing world

E. F. PATRICE JELLIFFE

Nutrition education, which preferably should be termed guidance, has been undertaken by a variety of personnel, often ill prepared to give relevant and practical advice, for several decades.

Although nutrition education has been stated to be a component of many nutrition programmes it has often been an ineffective tool for improving nutritional status at community level as the objectives of the programmes have been unclear and/or unrealistic and evaluation of their effectiveness has not been an integral part of the programme. Techniques used have often been unimaginative or poorly conceived and there has been a lack of well-trained or sufficiently motivated personnel to come to grips with the intricacies of guiding families or communities realistically in improving their dietary habits and life styles. The discontinuity of programmes and at times meagre funding have also contributed to a lack of success in this field.

Nutrition training

For nutrition guidance to be undertaken sufficient personnel from many disciplines must be trained in practical nutrition, which encompasses among others sound, realistic nutrition knowledge, the ability to communicate with different ethnic and socioeconomic groups, and cultural awareness of food and disease ideologies.

Nutrition trainers

These include trained professionals in the field of nutrition and bio-chemistry — e.g. research scientists, nutritionists, public health nutritionists, dieticians, home economists, and science and biology teachers. Professionals whose work includes some aspects of nutrition—e.g. food technologists, physicians, nurses who many have acquired further training in nutrition, dentists, agriculturalists, pharmacists, health educators, and community development workers—may also help in the training process. The assistance of medical or social anthropologists has been more actively sought in recent years. Unofficial nutrition educators, with no proven scientific basis for their teaching—e.g. food-fad cult leaders, herbalists, etc.—as well as unofficial change agents such as the food industry, through mass advertising of their products, have also entered the wide arena of nutrition, much to the confusion in nutritional knowledge of the lay public.

Nutrition training

Definition

This is the academic and practical instruction in nutrition, dietetics, and food science—usually including the scientific, economic, and social aspects of the subject.[1]

Recent surveys undertaken by two committees (V/1 and V/11) of Commission V of the International Union of Nutrition Sciences have revealed, on the whole, inadequate training in practical nutrition in schools of nursing. Teaching was not task oriented and the main trainers were nurses with an insufficient background in applied nutrition sciences.[2-4]

In schools of medicine surveyed in Europe, Asia, and Africa (Nigeria) it was noted that some basic nutrition was taught in departments of physiology and biochemistry. Training was undertaken by means of didactic lectures or in laboratory practicum. In 68 per cent of all schools physicians were in charge of nutrition training of medical students and the services of nutritionists or dietitians were little used.[1]

In the USA, since 1977, some improvement has been noted in this area,[5] and medical students are beginning to realize the importance of nutrition and have begun to request that the subject be included in a more relevant fashion in their curriculum.[6] Nutrition training in schools of dentistry and pharmacy included in the International Union of Nutritional Sciences Survey[1] also demonstrated a general lack of interest in the subject of human nutrition.

Fulop[7] has defined teaching 'as the process of helping learners to learn' and has commented that an individual, knowledgeable in his or her own subject, does not necessarily have the required skills to undertake this function adequately. It has been well recognized, for many years, that training methodologies in many fields have been irrelevant and inappropriate as the required skills to accomplish specific tasks by the students were not considered in curriculum planning. Clearly defined objectives must be established in terms of the tasks the students should be able to accomplish on completion of their training. These objectives have been classified in different categories of attitude, knowledge, and skills.[8-11]

Guilbert has also described a more specific taxonomy such as *institutional objectives* (e.g. within, for example, a centre for health sciences), *intermediate objectives* (developed within a department), and *instructional or specific objectives* (referring to objectives devised for a short learning period).[12] McGaghie *et al.*[13] have identified for any training programmes certain new educational roles for the trainer:

1 *The planner*—who will clearly define the objectives of the programme and would, for example, be able to identify the competencies in nutrition which the cadre of health professionals who are being instructed must achieve.

2 *The manager*—who will utilize available resources to achieve the desired results. When the training of health workers is considered the revision of curricula seems mandatory to permit maximum mastery of relevant nutrition knowledge and should also ensure a sustained and continuous exchange of ideas and information between student and trainer.

3 *The evaluator*—who will critically gauge whether the learning objectives and goals have been reached.

It is hoped that future trends in education will specifically emphasize the 'teaching/learning system'[14-16] rather than expecting, as was done in the past, that the spoken and written word from the educator would correctly impart practical knowledge and not merely memorized concepts fast forgotten after the final examination. The Center for Educational Development in Health (CEDH) has field tested a manual on the application of competency-based curriculum development techniques for health professionals in seven developing countries. Many issues were raised regarding the long-range effect of the competency-training programmes, such as the use of the newly acquired skills by the health workers, mechanisms required to institutionalize these programmes, as well as their cost and effectiveness. It was felt that the transfer of this new technology would have a limited effect unless it was accepted at the highest organizational level and a systematic follow-up instituted.[17]

A new look at nutrition training

No general model can be advocated to improve nutrition training among health cadres. The curriculum for each cadre of worker must be formulated according to the needs of each individual country as well as the available facilities and faculty within each institution.

Many guidelines and models have been formulated to improve the inadequate teaching of this important discipline among health personnel.[18,19] But the models or methodologies adopted must take into account the main nutrition problems of public health importance in the country and training must be directed at prevention, alleviation of the diseases, and nutritional surveillance.

As Guilbert has clearly stated: 'professional education needs to develop in students a broad and clear concept of the social role of his/her profession.'

A programme which is inefficiently taught by instructors ill versed in nutrition can hardly imbue the student with the enthusiasm, dedication, and sense of duty towards the community which is required. It must, however, be clearly imparted to health professionals that nutritional problems can be solved only by an interdisciplinary network of individuals who within their own curricula receive similar yet complementary instructions in nutrition, e.g. agricultural extension workers and community development personnel.

Curriculum changes

In many parts of the world curricula for all cadres of personnel appear to be subject centred rather than task or community oriented. A timorous approach is taken to the great effort needed to overcome the many difficulties and barriers which will be encountered when curricular changes are proposed.

Bryant has identified, among others, five obstacles to major changes in educational curricula: dynamic conservatism, institutional bureaucracy with built-in resistance to change, complexity of curriculum change, lack of institutional resources, and lack of a model upon which to base a curriculum change.[12] To these barriers could be added the psychological component, i.e. insecurity of trainers faced with the need to revise their mode of teaching or to be superseded in their work by new faculty if no facilities exist for continuing education or for re-education.

These barriers, however, may be penetrated in health-training complexes as long as the governing university, medical or nursing school administrative system is sympathetic and supportive to the long overdue change in curricula. In Third-World countries support from politicians may be obtained if the cost-benefit resulting from training health cadres in nutrition for specific tasks can be shown, for example in the diminution of severe nutritional syndromes and a lowering in morbidity and mortality rates in the population.

Representatives on curriculum committees need to be practical, experienced individuals in the fields of health and nutrition and knowledgeable regarding nutritional problems in the country. They must be aware of available facilities for training and of the resources and personnel which can be attracted to teach within the health complex. In some countries, for example, medical students, nurses, dentists, and dietitians attend together certain classes on nutrition, the same message is received, and multidisciplinary cooperation can begin at the classroom level. It is often important to have representatives of the student bodies on the planning of a new curriculum committee.

Various disciplines which interact within the field of nutrition and dietetics should be represented on the curriculum committee—e.g. public health nutritionists, physicians, nurses, dietitians, and sociologists as well as representatives from agriculture and other food-producing bodies in the country.

The principles for planning the curriculum must be carefully identified and in tune with any food and nutrition policy which may exist in the country. Practical applied nutrition, especially field experience (assessment nutritional status, home visits, task analysis, etc.) are an invaluable part of the teaching—learning process. The inclusion of an examination in nutrition may ensure that a course is actually given.

But examinations must endeavour to test levels of performance reached by the student congruent with the educational objectives set out in the new curriculum. It has been suggested that a new typology of test items be developed to make evaluation of the students' knowledge more precise.[19] A sound evaluation system must be built in to any new curriculum but more sensitive instruments must be devised which are meaningful and permit both students and trainers to interact and achieve better results and improve methods of training generally. It is now commonly believed that the use of a multidisciplinary team of individuals trained in different aspects of nutrition would be the ideal way to expose health cadres to the varied facets of nutrition—e.g., nutritionists, dietitians, biochemists, agriculturalists, and sociologists. Recent surveys have shown that nutritionists and dietitians have played a very minor role in the training of physicians and nurses.[1,3,4] This has been due in many instances to the unavailability of sufficient numbers of these cadres, or to a lack of cooperation and understanding—for example between nurses, doctors, and dietitians who do not appreciate sufficiently the complementary roles of their disciplines. In schools of medicine, it would be advisable when possible to have a chair of nutrition to give visibility to the discipline, permit students to communicate with the Nutrition Faculty during their training, and encourage recruitment into the field of nutrition. In schools of public health emphasis should be placed on the teaching of practical applied nutrition.

In schools of nursing, especially in Third-World countries, for the foreseeable future an insufficient number of trained nutrition experts will be available to teach nurses. It is therefore imperative that the training in nutrition of nurses tutors be upgraded, before their beginning their career as instructors of nurse trainees. Courses and refresher courses should be arranged in conjunction with training schools through existing agencies—e.g. nutrition institutes and international or bilateral organizations—and training should be equally geared as for other cadres to the relevant nutritional problems of the country.[2,4]

The need exists for books or pertinent training manuals in nutrition for health professionals who themselves are expected to be trainers in nutrition in primary health care programmes. Directors of nursing schools and tutors should, however, be aware of the availability of publications by the World Health Organization which will assist these cadres to be more effective generally, and help them evaluate their needs as well as their existing programmes in nutrition.[20,21]

Methods used in nutrition training should include some didactic instruction coupled with increased practical emphasis on group discussions, problem solving (e.g. real-life situations), role playing, simulation games (e.g. simulated nutrition systems SNS-FAO), field trips, the use of all available audio-visual aids, and evaluations by students of the use and efficacy of the local mass media.

Nutrition education

Definition

This is education of the public aiming at a general improvement of the nutritional status mainly through the promotion of adequate food habits, elimination of unsatisfactory dietary practices, introduction of better food hygiene, and more efficient use of food resources.[1]

Purpose of nutrition education programmes

The improvement of attitude, motivation, knowledge, and skills of the populations receiving this service. Nutrition education should be part of a general developmental programme and be channelled through all nutrition services —e.g. food production, processing, conservation, environmental hygiene, meal planning, intrafamilial food distribution, the careful selected use of food donations, meals for hospitalized patients, and school children and family planning programmes among others.

In the past, nutrition educators have assumed that the consumers of the service had little or no knowledge about nutrition, e.g. the 'empty vessel' concept, and in many instances because of their lack of training in applied communication, and insufficient knowledge regarding cultural habits of the population, many programmes were doomed to failure.

Fugelsang[22] has stated that applied communication is a 'creative function requiring both sensitivity and a creative ability' on the part of the educator. To these attributes could be added, curiosity in its widest and non-threatening context, a sense of humour, and compassion for those in need of assistance. Before undertaking any nutrition guidance programme, knowledge is required regarding food ideology, beliefs about disease causation, and taboos which may prohibit the use of nutritious foods by 'vulnerable' groups among other factors.

Food habits

Food habits have developed over time because of numerous factors, which include:

1 Physical factors, e.g. taste buds which encourage the acceptance or rejection of foods, organolepsis (mouthfeel) 'gluttony' (food gatherers and hunting groups, with no facilities for food preservation, e.g. 'mother nature's deep freeze').[23]

2 Ecological factors—food availability (game, roots, berries, domesticated animals, and wild grasses).
3 Knowledge of foods (e.g. avoidance of dangerous, poisonous plants).
4 Ease in food production (e.g. cassava versus millet).

Among traditional societies, both protective and harmful food habits developed over the years. Protective foods such as breast milk were needed for child survival. Foods eaten as part of a cohesive protective system, much dependent on the attitudinal behaviour of the society—e.g. avoidance if possible of a 'poor man's food', the use of prestige foods, adhering to religious rituals prohibiting partaking of certain animals (pork, beef, etc.)—would ensure spiritual protection. In many societies, male protection and survival was helped by providing them with more foods from animal protein sources, often to the detriment of the nutrition of mothers and children to whom these foods were declared 'taboo'. The acquiring of food habits is thus a deeply held aspect of the cultural past, including traditions handed on from the ancestors to the elders to the parents of children and their siblings, and encouraged by close relatives.[23] However, over the years, outside influences have eroded traditional food habits, e.g. improved communications, adverse weather conditions, natural or man-made disasters, and so-called 'change agents' from various fields of health and allied professions as well as unofficial change agents of the food industry, non-conformist groups, trend setters in the community, etc. (Fig. 10.1). These individuals have attempted to improve upon food habits, but at times some have had deleterious effects (e.g. bottle feeding, food fads).

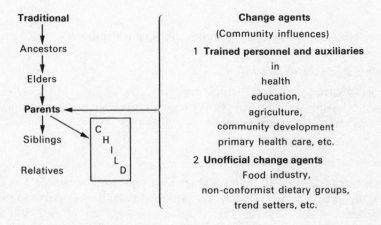

FIG. 10.1. A young child's world of food habits.

Thus a cultural frame of reference is required by nutrition educators; knowledge of existing food behaviour is imperative as well as the psychology of attitude which the educator will attempt to improve. Kagan[24] has defined an attitude as any belief which has an emotional component and believes that there are five main ways of acquiring attitudes. During childhood, attitudes are acquired by looking and listening, and rewards and punishments may be received if their attitude is felt to be appropriate or otherwise by parents or family members. As the child grows older, the need for a mechanism of identification arises, e.g. hero worship. During adolescence and adulthood, however, internal attitudes are acquired as a cultural definition of identity (e.g. caste, tribe, political viewpoint, etc.) develops. With maturity individuals will attempt to hold a consistent set of attitudes. The nutrition educator must be aware of the difficult hurdles ahead when a program is begun.

Communication

As already stated, the nutrition educator must realize that communication which occurs through the use of signs, symbols, words, and concepts is a two-way process.[22] This implies a dialogue with feedback, equality in the communication system, which has lacked in many nutrition education programmes as well as participation by the individual or larger audience. Many nutrition programmes have failed because of the lack of preparedness of the educator in this area. Multiple factors can cause a breakdown in communications, e.g. lack of practical knowledge of the educator, poorly prepared materials which are not culturally suitable, too little or too long a time devoted to the subject, lack of participation in the project by the audience, etc.

Characteristics of messages: communications considerations

Selection of media to be used
This will depend on many factors, e.g. cost, availability, suitability (e.g. literacy), possible outreach, physical resources, the training of educators in their use, and agency participation.

Characteristics of the messages[25]
1 Messages need to be attractive and credibly represented (valence).
2 Having identifiable objectives, stimulate audience participation (relevance).
3 Sensitivity to the culture, audience, message to be transmitted (congruence).

4 The message must be placed within a wider set of relevant assumptions (structure).

5 The message must be concrete (distinctive).

6 Thought must be carefully given to the positioning and spacing of the message, etc. (order and tempo).

7 The educator must be prepared for a lack of immediate impact of the message given to affect changes in attitudes, skills, and behaviours (realism).

Educational difficulties

In many communities, the work of nutrition educators is made difficult because of the various sources and agencies already working in this field of health, nutrition, population planning, agriculture, sanitation, resource management, etc.

The diversity of messages, range of actions, and competition among agencies may be a source of anxiety and frustration, unless some communication and cooperation is established among the groups.

Types of media

Media used may consist of traditional forms of communication, e.g. folk opera, dance troupes, puppet shows, or modern communication techniques (mass media such as texts, visuals, audio-media and audio-visuals). However, only face-to-face demonstrations stimulate the complete range of senses (Table 10.1).

TABLE 10.1   *Types of media*

I. Traditional communication forms
   Live performances: opera, dance troupes, puppet shows.
      Visual, auditory appeal, entertain   → learning

II. Modern communication forms: mass media
   1 Texts: books, magazines, flyers, pamphlets, comic books.
      Visual appeal, may entertain   → learning
   2 Visuals: photos, slides, film strips, photonovellas.
      Visual appeal, may entertain   → learning
   3 Audio media: radio and radio spots, tape recording, lectures, discussions, contests.
      Auditory appeal, may entertain   → learning
   4 Audio visuals: TV, sound movies.
      Visual, auditory appeal, may entertain   → learning

Face-to-face demonstrations stimulate complete range of senses

Characteristics of non-formal face-to-face nutrition education

This type of education has been the one most commonly used in the past. It is important to find a suitable time slot for the participants. Channels which have been commonly used have been the health services, community development, youth and women's programmes, and religious and civic organizations. Participants in these programmes should 'learn by doing'— e.g. prepare infant multimixes or their own 'nutripacks'—having grown or purchased suitable inexpensive culturally acceptable foods. Parents should be convinced that food alone has cured a malnourished child attending daily a nutrition rehabilition centre. Mothers should be able to monitor their child's progress by weighing and charting results on the 'Road to Health' chart, thus participating in community nutrition surveillance. Oral rehydration is another life-saving technique which can be used by village or town mothers. The difficulties with face-to-face, non-formal nutrition education is that it may reach at best only small audiences and be quite costly. However, cost-effectiveness analyses must take into account a number of variables, e.g. donated resources, the use of volunteers, the total cost associated with the programme, and numbers of participants who completed the programme and appeared to have been positively affected in their food behaviour.[26]

In all nutrition education programmes there is a need for built-in evaluation over time, including evaluation of the feasibility of the primary advice given (task analysis) and the effectiveness of the programme which is manifested by changes in attitudes, skills, and behaviour resulting in improved anthropometric measurements, lower mortality and morbidity among children, improved weight gains in pregnancy, etc.

Mass media and pictorial literacy

If mass media are to be used in nutrition education programmes, pictorial literacy must be determined. This is an informal process which will develop with long-term exposure to suitable, culturally integrated visual aids.[22] The interpretation of pictures will depend on experience and memory—e.g. constancy phenomena such as colour, shape, size, location, etc. Familiar cues which can be easily recognized must form part of the picture. Thus preparation of visual aids must take into account many features such as size, height, colour, perspective, shading, both the importance and danger of details, abstractions, etc. It is vital that in the preparation of mass media there should be a close team approach between the professional who has determined the content of the message and the artist who is concerned with the form of the message. All mass media should be carefully pre-tested among the different socioeconomic groups who will be exposed to this form of education.

Nutrition education programmes, progress, and trends

Programmes before the 1970s

Nutrition education in the early years had a limited scope and was mainly concerned with infant feeding associated with distribution of food supplements, advice to mothers in face-to-face demonstrations, and a limited use of some visual aids, e.g. posters and flannel graphs. Nutrition rehabilitation centres were also initiated[27] but evaluation of these programmes in the early stages was rarely undertaken. Nutrition education was given mainly through the health services channels, such as the MCH centres and young-child clinics; home visits were undertaken but their scope was limited by shortages of staff and transport. Preventive programmes consisted of immunizations —e.g. the BCG vaccination campaign in Kenya (1965) and the comprehensive Ankole Preschool Protection Program in Uganda in 1964—the under-five clinics, and the use of the Morley 'Road to Health' chart. In some areas nutrition education was provided through women's clubs and schools, the importance of school meals was stressed and horticultural projects were encountered, e.g. home and school gardens.[29] In the mid-1950s the Applied Nutrition Programmes were started through UNICEF and several evaluations of these projects were undertaken in countries such as the Philippines, Korea, India, Africa, and Latin and central American countries between 1961 and 1971.[30] Consultants in many instances agreed that projects had been poorly formulated and executed in the first place; sufficient resources and a sound infrastructure were lacking in many countries. A lack of cooperation was evident among the different minorities in these major rural development projects. Many common constraints were evident, e.g. a lack of trained personnel, staff, budget, logistical difficulties, etc. Many international staff were indaquately briefed, especially in relation to local culture and ecological conditions, and insufficient contact was made with their local counterparts.

Recent programmes after the 1970s

In recent years programmes have had a wider scope, as more emphasis has been placed on the complex causes of malnutrition and more research has been undertaken world-wide in this area. More attention has been paid to prevention of diseases, stressing the relationship between fertility control and food production—including processing and storage of commodities—the use and availability of water, and the need for widespread use of oral rehydration. The international attention now devoted to the need to promote breast feeding widely as well as the ill effects of some types of advertising and the monitoring of the activities of infant food companies are part of this change. More attention has been devoted to rural areas and the needs of farmers (seeds, fertilizers, irrigation, etc.) as well as emphasis on horticultural projects. Educational programmes for consumer groups have been much in demand, as in some countries these have become a more powerful force who want to take part in the shaping of more effective food policies. In the past decade the significance of commercial advertising on food habits of families has been much publicized. But these more comprehensive programmes are a slow outgrowth of the earlier ones which paved the way to these broader-spectrum endeavours.

These newer programmes have also reached a broader target audience in terms of numbers of individuals and groups. Changes in these areas include education in non-formal settings and more commonly with assistance from applied anthropologists. Methods used include face-to-face techniques, combined with the use of the mass media. Some evaluation has been undertaken of the effectiveness of mass media approaches, e.g. the use of commercial messages on radio, such as the USAID sponsored projects in which Manoff International and Synetics have provided their expertise and valuable contributions.

In 1974, a large-scale campaign using radio messages (a minimum of five 60-second spots per station for 15 months) resulted in significant increases in the intake of iodized salt among the target group (i.e. from 5 to 98 per cent).[31,32]

Other changes in knowledge and attitudes were reported in this campaign, e.g. in the area of breast feeding, intestinal parasites, the dangers of unsanitary drinking water, and protein-energy malnutrition. Another project, in the Philippines, using a radio spot campaign stressing to mothers the need to add oil to the infant's rice porridge (lugaw) as well as some fish and vegetables was also successful. This behaviour change was noted from the start of the campaign to the post-test surveys.[33]

However, a Harvard Institute for International Development (HIID) case study evaluation of this project found that the acceptance rate of mothers who had listened to the radio but had the message reinforced through a nutrition educator was 2.4 times higher than among mothers who had been exposed only to the radio message.[34]

Radio programmes have also been used in Tunisia (Dr Hakim programme) which at its inception included thirty-seven messages of 60-seconds duration for initially 4 months, on relevant nutrition topics to improve the nutrition of children and families generally. Experimental centres (with reinforcement of non-formal education) and control centres (no reinforcement of nutrition education) were selected. One of the conclusions which arose from the evaluation included the possibility that mothers who were exposed to both the radio message and non-formal education were more likely to have increased their knowledge of suitable foods to feed their family. Few mothers, however, reported changing their attitude or behaviour. The programme was continued with the addition of many new messages.[25]

More recent radio programmes include 2-minute nutrition spots from UNICEF in French and English which will be transmitted in a number of countries and was initiated in 1979. Similarly, the developing country-farm-radio network (DCFRN) includes radio tape messages distributed in sixty-five developing countries from Canada.

Other areas concerning *nutrition training* include:

1 Visual aids on xerophthalmia from the Helen Kéller, Inc. (USA), the

TALC SLIDE sets (Morley, UK), and the MEDDIA diagnostic sets for parasitic diseases from the Royal Institute of Tropical Medicine in the Netherlands.

2 Training manuals, such as the WHO Guidelines for Primary Health Care Workers (1981).

3 Training research such as the use of 'communication units' in the special nutrition programme by Gopaldas in India (1980).

Continuing education aids, from appropriate technology groups—e.g. 'Diarrhoea Dialogue' (from AHRTAG), General Nutrition (L.I.F.E.).

Conclusions

Many educators remain critical of the effect of nutrition programme on the general quality of life,[33] as they are generally dissatisfied with the lack of appropriate evaluation techniques, and the difficulties in obtaining cost-effectiveness information in the various projects.[35] However, it would appear that some prerequisites should be stated to ensure a maximally successful nutrition education programme:

1 The objectives should be clearly stated, i.e. behavioural changes expected, and numerical measurement of changes needed.

2 The teaching tools should be tested with the prospective audience.

3 The programme should be integrated into the overall development programme as reinforcement through multiple channels gives continuity.

4 Built-in evaluation must be an integral part of the programme as this allows for changes to be made, contributes to the improvement of the programme, and will measure cost-effectiveness.

From the available data in 1980, a consensus seems to have been reached:

1 Radio messages are most effective if they reinforce face-to-face programmes.

2 Multimedia approaches are usually more effective.

3 Information can be provided through many programmes.

In some, behavioural changes have been achieved, but few so far have been able to determine if permanent changes in food habits have been made. In this rapidly changing world in which political strife, inflationary prices, etc., are present, it may be likely that nutrition education messages may have to vary drastically in their messages to fit in to the more difficult socioeconomic conditions which surround us, and will have to be as scientific as possible, yet remain simple and essentially practical. However, if possible, multimedia studies should be undertaken with an evaluation of the most effective different combinations. Leadership and participation by qualified technical and research personnel should form part of the goal for which many have striven over the past years to allow all people to achieve their right to adequate daily nutrition.

References

1. Fidanza, F. Nutrition education and training in schools of medicine, pharmacy and dentistry, draft report. IUNS Committee V/1. 1978. Presented at *The International Nutrition Congress*, Rio de Janeiro, Brazil (1978).
2. International Union of Nutritional Sciences. Report of the 2nd Meeting of the IUNS Committee V/11. Nutrition Education in Nursing. *J. trop. Pediatr. Environ. Hlth.* **21**, 345 (1975).
3. Jelliffe, E. F. P. Nutrition in nursing curricula: historical perspectives and present day trends. *J. trop. Pediatr. Environ. Hlth.* **20**, 150 (1974).
4. Jelliffe, E. F. P. (ed.) The teaching of nutrition in schools of nurse auxiliaries. An international review of programmes. IUNS Committee V/II, IUNS Publication (1981).
5. Darby, W. J. The renaissance in nutrition education. *Nutr. Rev.* **35**, 33 (1977).
6. Jelliffe, E. F. P. *Nutrition education and training* (National Academy of Sciences) (1980). (Unpublished document.)
7. Fulop, T. Educating the educators. *WHO Chron.* **32**, 303 (1978).
8. Bloom, B. S. (ed.) *Taxonomy of educational objectives. The classification of educational goals. Handbook of cognitive domain.* McKay, New York (1956).
9. Krathwohl, D. R. *et al* (eds.) *Taxonomy of educational objectives. The classification of educational goals. Handbook II, affective domain.* McKay, New York (1964).
10. Mayer, R. F. *Preparing educational objectives*, 2nd ed. Fearon and Pitman, California (1974).
11. Davis, I. K. *Competency based learning technology, management and design.* McGraw-Hill, New York (1971).
12. Guilbert, J. J. *Educational handbook for health personnel.* WHO, Geneva, Offset Publication No. 35 (1977).
13. Megaghie, W. C., Miller, G. E., Sajid, A. W., and Telder, T. V. *Competency based curriculum development in medical education. An introduction.* WHO, Geneva, Public Health Papers No. 68 (1978).
14. McKenzie, L. *et al. Teaching and learning.* UNESCO, Paris (1970).
15. Segall, A., Vanderschmidt, H., Burglass, P., and Frostman, T. *Systematic course design for the health fields.* Wiley, New York (1975).
16. Katz, F. M. and Fulop, T. *Personnel for health care. Case studies of educational programmes.* WHO, Geneva (1978).
17. Vanderschmidt, L., Massey, J. A., Arias, J., Duong, T., Haddad, J. Noche, L. K., Kronfol, N., Lo, E. K. C., Rizyal, S. B., Shreshta, M. P., and Yepes, F. Competency-based training of health professions teachers in seven developing countries. *Am. J. publ. Hlth.* **69**, 585 (1975).
18. Frankle, R. T. *Nutrition education in medical schools. A curriculum design.* Nutrition Foundation, Washington, DC (1976).
19. Downie, N. M. *Fundamentals of measurements. Techniques and practices.* University Press, New York (1967).
20. Wakeford, R. E. *Teaching for effective learning. A short guide for teachers of health auxiliaries.* WHO, Geneva (1974).
21. Allen, M. *Evaluation of educational programmes in nursing.* WHO, Geneva (1977).

22. Fugelsang, A. *Applied communications in developing countries. Ideas and observations.* Dag Hammarskjold Foundation (1973).

23. Jelliffe, D. B. and Jelliffe, E. F. P. Food habits and taboos: how have they protected man in his evolution? In *Progress in human nutrition* (eds. S. Margen and R. A. Ogar) Vol. 2 Avi Publishing, Conn., p. 67 (1978).

24. Kagan, J. The psychology of attitudes. *Forum* (on Attitudes, Behaviour and Human Potential), 4 (1973).

25. Munger, S. J. *Mass media and non-formal nutrition education* (A.I.D. Contract/To. C. 1198). Synetics Corporation, October (1978).

26. US Agency for International Development. Office of Nutrition. *A field guide for evaluation of nutrition education. An experimental approach to determination of effects on food behaviour in lesser developed countries.* US AID, Washington, DC (1975).

27. Research Corporation. *A practical guide to combatting malnutrition in the preschool child. Nutrition rehabilitation through maternal education.* Appleton-Century-Crofts, Meredith Corporation, New York (1969).

28. Cook, R. and Jelliffe, D. B. (eds.). Recent experience in maternal and child health in East Africa. Monograph 2. *J. trop. Pediatr. Afr. Child Hlth.* **12**, 3 (1966).

29. Jelliffe, D. B. Infant nutrition in the subtropics and tropics. In *Nutrition Education. General Considerations.* 2nd ed. WHO, Geneva, WHO Monograph Series No. 29 p. 217 (1968).

30. McNaughton, J. W. *UNICEF and child Nutrition in the seventies and beyond. An assessment in priorities in child nutrition,* vol. III. (Health Services and Education in Relation to Nutrition.) Harvard University School of Public Health, p. 375 (1975).

31. Cooke, T. M. Using radio for nutrition education, part II. In *Nutrition planning in the developing world.* CARE, p. 83 (1976).

32. Manoff, R. K. A summary of the mass media nutrition education project in Ecuador (1974–75). *PAG Bull.* **6**, 27 (1976).

33. Rasmusen, M. *Current practice of future direction of nutrition education in developing countries. A research and policy assessment.* Nutrition Planning Information Service, Ann Arbor, Mi., Document No. 181 (1977).

34. Austin, J., Anderson, M. A., Goldman, R., Heimendinger, J., Overholt, C., Pyle, D., Rogers, B., Wray, J., and Zeitlin, M. *Nutrition intervention assessment and guildelines.* HIID (1978).

35. US Agency for International Development. Office of Nutrition. *Application of a field guide for evaluation of nutrition education in three programs in Brazil. An experimental approach to determination of effects of food behaviour in lesser developed countries.* US AID, Washington, DC (1976).

11 The TALC experience

DAVID MORLEY AND FELICITY SAVAGE

The beginnings of TALC

Teaching Aids at Low Cost (TALC) began in the London School of Hygiene and Tropical Medicine in the middle of the 1960s. Its aim in those days was to provide slides for nutrition students who were going overseas to use for their teaching. But the time required to sort out and copy individual slides, and the cost of doing so, made it necessary to devise and produce standard sets. We started from the maxim that any picture is worth a thousand words, that a colour picture is worth more; and an actual photograph of what you are talking about, yet more again. This is especially true in medicine and allied subjects, in which students have to learn to recognize appearances.

Accepting that, it became clear that the transparency is the most economic kind of visual teaching material. The economy is maximal if the transparencies are produced in standard sets of twenty-four horizontal frames, and sent out in the form of a strip with self-sealing mounts for users to mount themselves. We believed from the outset that such a set of slides, accompanied by a detailed text (Fig. 11.1) could be a great help to a range of teachers and students. This view has been vindicated by TALC's sale of over two and a half million transparencies in the 15 years of its existence, and we know from our researches that each slide is seen by an average of eighty students in the first year after it is received.

FIG. 11.1. As so many teachers in developing countries have to pay for their own visual aids, we send a high proportion of our material in this low-cost form, leaving the mounting to be done overseas. Sent in this way the cost, including postage and packing, is about half the cost of transparencies the teacher might take.

About fifty sets of twenty-four slides are available—some linked in pairs, and one 'ten set' collection of 240 slides. They are mostly about mother and child health, or MCH, and nutrition. These are vital subjects of great national importance everywhere; yet in teaching hospitals, where the majority of health workers are trained, they are not prestigious, and have low priority. If 'MCH' and nutrition are to command students' interest and attention as they deserve, good teaching aids must be available. It is striking how the world-wide demand for material such as TALC distributes has grown at the same time as interest in these subjects has grown (Fig. 11.2).

FIG. 11.2. This excellent logo was created by one of our part-time workers.

Unusual features

During the 15 years of its existence TALC has prospered and grown without any formal offices. It is run by a group of women from their own homes. Perhaps in this, TALC is looking into the twenty-first century when, with instant reliable telecommunication, more people will work from their homes. The efficiency of our organization should be a challenge to businessmen who work from offices. There is only one full-time worker—a mother with three children who uses a small office off her kitchen. She directs some fifteen other housewives who work part-time packing up slide sets and books and posting them, mostly through their own local Post Offices to addresses all around the world. TALC has made good use of the enormous untapped potential of married women who are very willing to undertake a semi-skilled job that they can do in their homes. They are paid 'piece work' and what they receive amounts to slightly less than a fully-trained secretary can earn by taking in typing. But, above all, TALC has been great fun to everybody involved. We are proud of our logo designed by one of our workers and now known world wide (Fig. 11.2).

Communicating new ideas

TALC has made it possible for the Tropical Child Health Unit in London to spread a number of new ideas extensively around the world. One of these is the children's growth chart—sometimes called the 'Road to Health Chart'—which was first developed in the late 1950s in west Africa. This chart (Fig. 11.3) is a means by which all health workers can participate in promoting adequate growth. Many child health specialists now see promoting growth as a much more appropriate objective than treating, or even preventing, malnutrition. TALC makes charts available in five languages, and there are a number of easily used teaching aids to go with them. Growth charts of the same general type are now used world wide. There are many local adaptations, and in a few countries they have become national documents.[1]

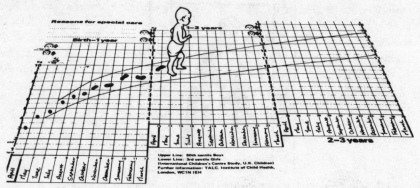

FIG. 11.3. TALC has played a major part in these simple growth charts, now used in almost every developing country.

But growth charts and the primary child-care system that they are part of cannot be accepted, understood, and implemented overnight. In places they have not yet reached, an even simpler technique has proved useful, which makes it possible to assess the nutritional status of a young child instantly. This is the measurement of the mid-upper-arm circumference (or MUAC). The technique reaches the peak of streamlined simplicity and virtual freedom from cost in the 'Shakir Strip'—after Professor Adnan Shakir of Baghdad who developed it. You make your own strip from any material that is reasonably strong, and which does not stretch—anything from X-ray film to banana fibre. You select your cut-off points (usually 12.5 and 13.5 cm—though they may vary slightly in different populations) and mark the strip to show the range of the circumferences of well-nourished, mild, and severely malnourished arms. We can then use the same strip to measure the upper arms of any children between the ages of 1 and 5 years. We can identify quite accurately their state of nutrition without knowing their exact age, because during those four special years even a well-nourished child's arm circumference changes little. The use of this strip since its first description in 1975 has spread rapidly (and to a large extent through the TALC network) and is known, not only by many health workers, but also by many workers in other fields such as agriculture.

Feeling for malnutrition

There is now a further modification and simplification of arm-circumference measurement. Experience has shown that both health and lay workers can rapidly learn the feel of different children's arms. All that is required is a bag of 'model arms' of different diameters made of pieces of wood or rolled-up newspaper (Fig. 11.4). Workers put a hand in the bag and learn to select blind the 'arms' with a circumference of less than 13.5 or 12.5 cm. Once the feel of cylinders of these sizes has been learnt, malnourished children can be confidently identified by feel of the arms. It is not entirely subjective, because they learn where on the thumb the nail of the first finger reaches, at whatever is chosen as the local 'cut-off' point. This simple technique, first suggested by Laugesen in India, may cause a new greeting for children to spread through developing countries. Perhaps a new concept of the teaching of nutrition may develop which will depend on feeling children's arms and discussing with parents and the community what tissues make up an arm and how it can be strengthened.

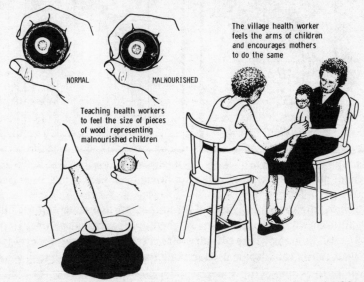

NORMAL   MALNOURISHED

The village health worker feels the arms of children and encourages mothers to do the same

Teaching health workers to feel the size of pieces of wood representing malnourished children

Fig. 11.4. This further development from the Shakir strip comprises health workers calibrating their fingers and thumbs in such a way that they can measure an arm circumference by 'feel'.

The Nabarro wall chart

A child who is particularly at risk is the wasted child, who may have grown well in the past, but who has now lost weight. Collectively, these children are known to have a high mortality. To identify them, the arm circumference has several advantages over a single weight-for-age measurement—especially when a child's age is not known accurately. A strong competitor for this purpose has always been the measurement of the child's weight for his height —but until recently the only known techniques for assessing this ratio were too complicated for field use, and the need for a double measurement multiplied any inaccuracies. However, in 1980, Nabarro, with help from the Save the Children Fund, designed a wall chart which enables us to evaluate weight for height while only measuring weight. A prototype of the chart was first used in Nepal. Children (whose ages are not necessarily known) are weighed, and then put to stand against the chart on the wall. They stand in front of the column for their weight. If a child's weight is appropriate for his height, the top of his head reaches a green band (Fig. 11.5). If, however, a child is wasted, his weight is appropriate in health only for a shorter child. So, he is too tall for his weight and the top of his head reaches a higher red band.[2]

 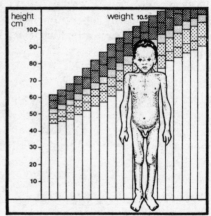

FIG. 11.5. TALC has helped in the wide distribution of these wall charts, which make assessment of nutrition in terms of weight for height quite simple.

These charts are particularly useful for immediate assessment of the nutrition of children who have not been weighed and measured before. Hence, they are useful in areas where primary care services are not well developed. But the wall charts can also be used in clinics alongside the ordinary growth chart, where they are useful to assess older children attending for the first time. Also, a child whose head reaches the red band can have a red circle put around the dot on his weight chart—a warning to health workers, even though the child's weight may not yet be below the lower line. The wall chart can also save embarrassment to mothers with children who, for instance, are small at birth and grow poorly in their first 6 months or year, or who are genetically small. Such children are likely to remain small and, despite adequate feeding later, their growth curve may stay below the lower line on the growth chart. Unfortunately, health workers may tell these mothers again and again that they are failing to feed their children adequately—thus hurting the mother and sacrificing their own credibility. However, if a wall chart is available, we can measure the child against it, then, if his head reaches the green, we put a green circle around the dot on his weight chart, and everyone will know that he is adequately nourished for his height, even though his growth curve appears to be too low. Through TALC we expect these charts to be widely distributed, and they can add much to the ability of health workers to correctly assess the nutrition of children. This is an example of the valuable role that the organization can play in hastening the spread of useful new ideas.

The salt and sugar spoon

TALC has also helped to spread innovations in other fields. Perhaps the most important of these is the Hendrata Spoon, after Dr Lukas Hendrata from Jakarta. Towards the end of the 1970s, Indonesia led the way in the development and national implementation of more appropriate ways to rehydrate children with diarrhoea, both in hospital and at home. Some groups in Indonesia were particularly concerned with village health workers. For these Hendrata[3] developed a simple measuring spoon for salt and sugar. It is intended for mothers to use at home to make a solution suitable to give to children soon after a first diarrhoeal stool—therefore before any dehydration is possible. A special feature was that the instructions about how to use the spoon were stamped into the plastic. This idea has been further developed by TALC, and with OXFAM's help a special mould was produced on which it was possible to change the languages without having to make a new mould (Fig. 11.6). Currently these spoons are available with instructions in Arabic, English, French, Portuguese, and Spanish. We hope that eventually every village health worker and perhaps every school teacher in the villages of developing countries will hold such a spoon or a locally made wooden measure. Some believe that this technique of giving oral fluids can do as much to save life as did the introduction of penicillin in the 1940s. That TALC can help to get an idea from Indonesia to Africa, or from Nepal to the Americas depends, very much on the two-way nature of its communication network. People working in the field wanted to share their ideas—and allowed us to help them do so.

A typical month's work

Although spreading new ideas is an important function of TALC, possibly the day-to-day work of sending out teaching aids and books still has, overall, a greater influence on mother and child health in developing countries. In any one month the organization receives altogether about 200 requests for

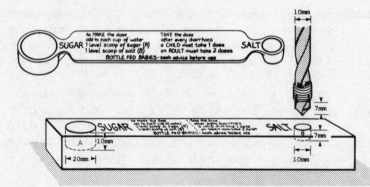

FIG. 11.6. Overcoming dehydration by oral rehydration through the 1980s will save hundreds of thousands of lives. These double-ended spoons distributed by TALC, or the locally made wooden measures, are one practical step by which this can be achieved.

material. In response to these requests, the housewives have to send out some 25 000 transparencies arranged in their sets of twenty-four with their text. In the same month they are likely to send out 1700 books, perhaps a quarter of which are chosen from the two or three most popular titles. A particular favourite is David Werner's book *Where There Is No doctor*, of which TALC alone despatches an average of 500 copies a month. By the end of 1980, TALC will have despatched some 10 000 copies of this book, their total weight being over 7 tons.

During the month we shall expect to pay out US$3200 in wages to our part-time workers, but these make up only 15 per cent of our costs. We will also expect to receive around US$15 000 in payment for goods received, mostly in small sums by cheque, all to be patiently handled by our hard-working bookkeeper.

For whom TALC material is intended

Most of the material supplied by TALC is for use in training schools—for medical auxiliaries, nurses, and medical students. The materials are neither suitable nor intended for the general public at village level. Visual aids for the village need to be specific to the area, so they usually need to be produced locally. A possible exception to this is the flannelgraph, and recently—thanks to an enormous task undertaken by Gill and Sue Gordon—an excellent flannelgraph has been produced which may be appropriate for use in large areas of Africa. The flannelgraph lends itself to audience participation and is particularly appropriate for small group teaching at village level. You place the flannel cut-outs of the subject to be taught on a flannel background to which they cling—and so you build up a picture along with the story that you are telling.

Disseminating improved ideas on teaching

We have always been concerned to find ways of developing the educational value and efficiency in use of both slides and the texts that go with them. Over the years we have found that it is valuable to build up a slide from two or more transparencies. Often only part of an original transparency is required to demonstrate an idea, and the composition of the picture is almost always improved when it is trimmed. But an even greater benefit is that the other half of the frame can be used to show something else that is relevant, and their apposition may help to link the appearances and the ideas they convey in the student's mind. One half of a slide might show a leg with some lesion, and the other half the X-ray of the leg. We can show the distribution of a rash on a child's whole body on one half of a slide, and a detailed close-up of the spots on the other. We can show a child on one half and his weight chart on the other. Or we can use the two parts of the slide to compare different conditions that confuse by their similarity—such as a fungal patch and indeterminate leprosy. This way of linking ideas in a student's mind is, we believe, a useful and underused technique in teaching.

From the early days of TALC we have always labelled slides with a code letter and number so that even if the mount is not marked, the set from which the slide arises and its position in that set can be identified easily. This is also a help when you are reading a script about a projected slide—you can see on the screen which one you should be talking about, without dismantling the equipment. We have also learned the value of putting 'Letraset' arrows and numbers on slides, to direct student's attention to important features. We have learnt how confusing 'right' and 'left' can be even to the most sophisticated audience—and here again, letters on the slides can overcome the problem.

Simpler but better English

Perhaps the most important development has been a new approach to the preparation of texts and tapes.[4] Only recently has it become widely appreciated that health workers (including doctors), whose first language is not English, may not be able to follow a text easily. These students may be socially fluent in the language—but they have more difficulty with reading and with understanding technical material than we realize. Even students who enter training school after 4 years of secondary education in English may have considerable difficulty. And there are many with far less experience of the language than that. Educationalists have identified a number of common problems, and recommend a specially controlled language form with short sentences, restricted vocabulary, and simple syntax. The result can, if prepared with care, sound deceptively ordinary to the native Anglophone. But for everyone else it is strikingly easier both to comprehend and to translate accurately.

However, it is not only the language itself that can be confusing. The arrangement of the ideas in a text; the amount of detail given; the extent to which elaborate concepts are broken down to their component parts and whether the material is presented in a personalized, situational form, or a generalized abstract one, all greatly influence the amount of learning that a slide can effect.

Recently, one of us (F.S.), working in a provincial area of a developing country has been able to test slides, texts, and tapes with several groups of community-nurse students. Above all, let it be said that these students and their teachers enormously appreciate the most meagre material—more than justifying the whole endeavour. However, a number of problems and some solutions to them have been identified, so that new and revised material should be both easier to use and more effective. For example, an important textual point must be visually illustrated if students are to grasp it. You cannot 'hang' a point on a picture that does not directly illustrate it, even though the two may be related.

Detailed description is appreciated—and, provided the audience can see what you are talking about, length is well tolerated. But if the script then becomes too long, it is necessary to divide the set, and not attempt a session that continues for more than 1 hour. Most important of all, the students' enjoyment and learning are both much greater if they can stop and discuss slides, and identify the appearances referred to for themselves. So we have adopted a 'Question and Answer' format for the texts, which stimulates the audience to look and to think. The experience overall confirms that there are a number of subjects which people can learn more easily from slides than from a blackboard and books; we consider it is important to develop slides to do their particular job effectively, without trying to make them substitute textbooks.

TALC and the future

The first 15 years have opened our eyes to the enormous desire that health workers and their teachers in developing countries have for the teaching aids and books made available by TALC. Yet we know that we are producing only a minute fraction of the vast mass of material that could be required to reach WHO's goal of 'health for all by the year 2000'. We believe that 'small is beautiful' and that far better than TALC expanding indefinitely to try to meet this enormous need would be for small sister organizations to start all over the world in both developed and developing countries. Already some exist. The Voluntary Health Association of India (VHAI*) has produced much high-quality material of its own, particularly appropriate for use in India. At the same time it has been able to take items from TALC, to modify them slightly where necessary, and to make them available for purchase locally with rupees.

* C14 Community Centre, Safdarjang Development Area, New Delhi 110 016, India.

For many countries, in order to use the slide sets it is necessary to translate the written text into the local language and idiom—and sometimes to change words on the transparencies. This nearly always needs to be done in the country where the slides are to be used, and it is certainly a major task. But it is generally much easier to translate than to start again from the beginning—because collecting together the right twenty-four slides to teach an appropriate message, working out the text, and preparing them all for printing can take years.

It is perhaps one of the useful things that interested people in industrialized countries can contribute: they can collate valuable material from a wide variety of sometimes inaccessible sources and make them into teaching aids readily available at low cost to anybody anywhere in the world—for them to use, translate, and adapt according to their capacity to do so and their needs. This is the objective of TALC.

References

1. Morley, D. and Woodland, M. *See how they grow*, Macmillan Press, London (1979).
2. Nabarro, D. and McNab, S. A simple new technique for identifying thin children. *J. trop. Med. Hyg.* **83**, 21–33 (1960).
3. Hendrata, L. Spoons for making glucose–salt solution. *Lancet* **1**, 612 (1978).
4. Savage, F. and Godwin, P. Controlling your language: making English clear. *Trans. R. Soc. trop. Med. Hyg.* **75**, no. 4 (1981).

12 Economic and business aspects of infant formula promotion: implications for health professionals

PIERRE A. BORGOLTZ

Introduction

Industry assistance to the medical profession can certainly be beneficial when it contributes to the achievement of health objectives. Business strategies have a logic of their own, however, and business objectives do not always coincide with those of the health profession. Exploring the interactions between industry and the health professional has been avoided as a 'taboo zone', fraught with risk of controversy.[1] A better understanding of industry dynamics and operations could greatly benefit health professionals who are not aware of the factors coming into play in the formulation of business strategies, especially those that directly influence health practice.

This chapter outlines some of the basic considerations underlying industy behaviour in the infant-formula industry. Most important is the effect of infant-formula marketing strategy, and the exent to which breast-milk substitute producers have come to influence attitudes among health professionals regarding infant-feeding methods. Relevant economic and business information also includes the nature and strength of competitive pressures in the industry, and how private enterprise responds to specific market characteristics and structure.

To assess the effect of industry promotional practice, it is necessary, first, to see the infant milk industry as transnational in character and highly concentrated in structure. The business outlook for the industry in its traditional major markets is presently limited. The market positions of leading firms are well entrenched there, and the stage is set for an intensification of competitive pressures to keep up sales growth rate.[2]

The relatively untapped markets of developing countries offer industry the potential for future expansion. Consequently, the escalation of promotional activities that build and maintain the backbone of higher market share and a secure position of market power is to be expected in developing areas.

The thrust of these promotional activities is the situation in which infant feeding is initiated, in the maternity hospitals, and towards doctors and other health personnel who have a direct demand creation effect. Thus, infant-milk* industry promotion becomes critical in the adoption of artificial feeding and consumption of breast-milk substitutes.

---

* In this chapter, infant milk is used as equivalent to breast-milk substitutes, e.g. any food being marketed or otherwise represented as a partial or total replacement for breast milk,

From a business point of view, certain promotional effects are essential to generate brand product sales, as well as to expand the breast-milk substitutes market. These efforts, by definition an undermining of the practice of breast feeding, can only pave the way for problems in public health.

Close scrutiny of marketing practice for breast-milk substitutes and the industry's interaction with the health profession moves swiftly into a forbidden territory; specifically, the issue of what is appropriate public control of these practices. The continued demise of breast feeding is, and should be, a serious concern of the health profession.[3,4] Despite industry claims to the contrary, it is difficult to see how strong regulations and control of industry marketing practices can be avoided if breast feeding is to receive genuine endorsement and support in both developed and developing countries. Moreover, recent disclosures show that leading firms in breast-milk substitutes production have undertaken extensive and deliberate efforts to undermine the legitimate processes of the health and scientific community to resolve the issue on the basis of mutual understanding.[4-6] Considering the nature of the firms' breast-milk substitutes business and the extent to which they are willing to go to advance their narrowly defined corporate objectives, it is clear now that little lasting progress will be possible in the way of improving infant-feeding practices short of mandatory controls over the industry and its marketing operations.

The transnational infant-milk industry

Structure of the world-wide infant-milk market and leading firms

The infant-formula market was valued at an estimated US$2000 million world wide in 1979. The USA alone accounted for a quarter of world sales, or about US$500 million in 1979.[2,8] Sales in developing countries have been fastest growing during the last decade, and their share rose from one-third to one-half world sales by 1980.[9,10] Major processing facilities are located in developed countries. In most developing countries, infant-milk products are imported, packaged or in bulk so that local production is limited to reconstitution of the imported ingredients and packaging.

World wide, only about thirty processors are engaged in the production of infant formula. (Table 12.1.) This small group is dominated by a handful of very large, diversified firms. The four largest, with over US$240 million breast-milk substitutes sales, each account for an estimated 75 per cent of world industry revenues, including over two-thirds of the developing countries' market.

Nestlé (Switzerland), the product sector leader, holds fully one-third of world-wide sales, and over 45 per cent of the developing countries' market.[2] (Table 12.2.) American Home Products (USA) is second in the developing countries market, with a 12 per cent world market share.

whether or not suitable for that purpose—[90] including infant formula, full-cream powdered milk, evaporated and condensed milk.

TABLE 12.1　*Leading transnational companies of breast-milk substitutes: total and foreign sales, 1979*

Rank	Company	Country of origin	Total sales 1979 (US$ million)	Percentage of foreign sales	Breast-milk substitute sales
		Milk companies			
1	Nestlé	Switzerland	14 113	96	700
2	Borden	USA	4313	20	NA
3	BSN-Gervais Danene	France	3866	40	40†
4	American Home Products	USA	3649	34	240
5	Foremost-McKesson	USA	3313	NA	NA
6	Snow Brands	Japan	3239	—	15†
7	Carnation‡	USA	2826	33	NA
8	Bristol Myers	USA	2753	31	250
9	Sandoz	Switzerland	2673	97	NA
10	Unigate	UK	2202	12	50†
11	CDC	Canada	2015	70	25†
12	Abbott Laboratories	USA	1683	36	335
13	Meiji	Japan	1438	1	50†
14	Morinaga	Japan	1227	1	65†
15	Wessanen K.	Netherlands	1319	57	NA
16	Glaxo	UK	1080	61	90†
17	VARTA	German Federal Republic	900†	46	125†
18	D.M.V.	Netherlands	810†	70	NA
19	Syntex	USA	471	45	15†
20	Domo-Bedum	Netherlands	415†	NA	NA
21	Frico–CCF	Netherlands	370‡	10	NA
22	Nutricia	Netherlands	230‡	40	65†
		Infant foods companies			
1	Unilever	Netherlands–Great Britain	21 744	NA	
2	Heinz	USA	2471	38	
3	Beecham	UK	1792	67	
4	Reckitt Coleman	UK	1399	77	
5	Gerber	USA	500	20	

* NA = not available. Including evaporated or powdered milk marketed as infant milk.
† 1978.
‡ 1977.
Source: The Foreign 500. *Fortune* Aug. 11. 188 (1980).

 The second largest processor, Abbott Laboratories (USA), with a 16.5 per cent world market share, is the dominant firm in North America, accounting for more than half of sales. Bristol Myers (USA), with a 12.5 per cent world-market share, is the second largest in North America.

 The high level of world-market concentration and dominance by the four largest processors corresponds to an even higher level of market concentration in individual country's markets. The market power of the leading firms is illustrated in Table 12.3, which details rank and/or market shares in a sample of twenty-four countries. A very highly concentrated structure of the infant-formula industry is observed in all these countries, with the same three

TABLE 12.2   *Infant formula sales—leading transnational corporations, 1979*

Firms	Total sales (US$ million)	US sales (US$ million)	Foreign sales (US$ million) (% total sales)	Sales in developing countries (US$ million) (% total sales)
Nestlé	700/650	–	675 (97)	400 (57)
Abbott Laboratories	335	225	80 (25)	40 (12)
Bristol-Myers	250	160	90 (36)	70 (28)
American Home Products	240	35	205 (85)	95 (40)
Grand total	1525	455	1045 (68)	605 (40)

Source: Based on annual reports, trade sources and estimates from hearings. Sub-committee on health and scientific research. United States Senate, 95th Congress. 2nd Session. Examination of the marketing and promotion of infant formula in developing nations. May 23. Washington, DC (1978).

or four companies controlling at least 90 per cent of the product market in the nine developed countries and sixteen developing countries for which data are available. The four leading firms are present in almost every market, with one of the leaders at least enjoying a dominant market position in all markets except Japan. Nestlé is ranked first, and enjoys at least 60 per cent market share in four of the nine developed countries, and in fourteen of the fifteen developing countries. American Home Products ranks second in nine of the fifteen developing countries, with market shares ranging from 10 to 40 per cent.

The transnational expansion of the leading infant-milk processors is associated with a high degree of foreign dominance in national markets. In all but five of the countries, foreign companies account for over 55 per cent of national markets. Local firms control their domestic market only in Japan and the USA. In the Federal Republic of Germany, the Netherlands, and the UK, the market share of foreign firms ranges from 25 to 35 per cent. In developing countries, denationalization of domestic market is complete, reaching 100 per cent in all cases, except for Algeria, Mexico, and Sri Lanka, where State-owned enterprises have been set up, and in India (see Table 12.3). Tables 12.1–12.3 show the infant-formula industry to be characterized world wide by a high concentration among a few very large corporations of trans-national reach.

Leading infant-milk processors. Another important consideration with dynamics and market behaviour in the infant-milk industry is that all but a few small infant-milk producers are large, diversified food or pharmaceutical transnational corporations based in the developed countries. Sixteen firms among the twenty largest infant-milk producers, which account for over 95 per cent of world output, each had total sales exceeding US$1000 million in 1979 (Table 12.1). By and large, leaders in the world infant-milk industry are highly diversified firms, with 92–98 per cent of their revenues accruing from other product lines. Geographic expansion constitutes another important growth option for these firms' business strategy. Three-quarters of the developed countries' infant-milk producers derive more than one-third of their total revenues from their foreign operations, the major exception being the Japanese companies.

The transnational infant milk leaders were characterized by high rates of sales growth, particularly in foreign operations. These firms engage in advanced business and marketing techniques to attain their common cor-porate objectives of profit maximization through increased sales turnover, increased market shares, and rapid introduction of new products on a world-wide basis.[11] They tend to operate in concentrated products markets, where they enjoy significant market shares. Large resources are devoted to adver-tising and promotion of their branded products. Advertising-to-sales ratios are high, ranging from 10 to 15 per cent in their pharmaceutical sales,

TABLE 12.3   *Concentration of infant formula markets in selected countries. Late 1970s*

Country	top 1	top 2	top 3	top 4	Year	(US$) million	Foreign firms (% market share)
	\(% cumulative market share and company)					Market value of sales	
Denmark	60(N)	90(D)	100(M)		1976	NA	100
France	65(N)	95(BSN)	—		1978	110	65
Italy	55(N)	95(M)	—		1979		95
USA	55(A)	90(BM)	100(W)		1979	500	0
Canada	60(A)	60(BM)	100(W)		1979	—	100
UK	40(CG)	75(G)	99(W)		1973	40	25
Japan	40(Mi)	80(Mo)	90(S)		1973	NA	9
West Germany	50(M)	85(N)	95(H)		1977	125	35
Sweden	50(N)	100(Local)	—		1979	—	50
India	35(Amul)	51(G)	65(N)	77(N)	1975	51	65
Malaysia	40(N)	80(CCF)	90(D)		1977	20	100
Hong Kong	NA	NA	NA		1978	17	100
Sri Lanka	32(N)	58(Local)	83(G)	94(G)	1978	20	75
Thailand	70(N)	—(W)	—(D)	—(BM)	1976	20	100
Philippines	47(N)	85(W)	99(BM)		1975	22	100
Ethiopia	40(N)	65(Mo)	85(Mi)	95(A)	1975	1.5	100
Nigeria	40(N)	59(W)	77(CG)	90(A)	1975	50	100
Tanzania	80(N)	89(W)	98(A)		1977	2.5	100
Columbia	70(N)	97(W)	100(A)		1976	20	100
Peru	65(N)	90(W)	98(A)		1976	10	100
Mexico	55(N)	30(Local)	—(BM)	95(W)	1978	50	70
Venezuela	50(N)	80(W)	95(BM)		1978	NA	100
Brazil	95(N)	—	—		1978	70	100
South Africa	72(N)	90(W)	—		1979	65	100
*Panama	75(N)	87(A)	95(BM)	—	1978	22	100

NA = Not available.
* = Imports.
Source: United Nations Center on Transnational Corporations. Transnational Corporations in Food and Beverage Processing, New York. ST—CTC—19 (1981): trade journals.
Companies: N = Nestlé; D = Dumex; BSN = BSN–Gervais Danone; A = Abbott Laboratories; BM = Bristol Myers; W = Wyeth Laboratories; CG = Cow & Gate; CCF = Frico; G = Glaxo; Mo = Morinaja; Mi = Meiji; M = Milupa; H = Humanische Milch.

including infant nutritionals.[12] Returns on investments for these products are equally high, with profit margins commonly exceeding 15 per cent and ranging up to 25 per cent (Table 12.4).[12]

Patterns of competition in the infant-milk industry

The high levels of market concentration that characterize the infant-milk industry have been associated with oligopolistic market behaviour. Non-price competition, emphasis on product differentiation, and high promotional expenditures predominate in the business behaviour of major firms.[13] Classic economic theory argues that leading firms tend to avoid price competition, either through collusion or through tacit understanding that price-cuts by any of them would affect each other's market shares and lead to

TABLE 12.4 *Profit margins, advertising (R & D) research and development of Pharmaceutical operations: leading infant milk firms (including breast-milk substitutes), 1978 (as percent of total pharmaceutical sales)*

Firms	Pharmaceutical operations including breast-milk substitutes		
	Profit margin	Advertising	R & D
Abbott Laboratories	20.0	11.5	7.5
American Home Products	24.0	10.0	3.3
Bristol—Myers	18.0	15.0	4.0
Nestlé*	13.0	10.0	5.0
Sandoz/Wander	17.0	14.5	12.9
Syntex	25.0	15.0	10.0

Source: Reference 12.
* Only pharmaceuticals.

devastating price wars. Price wars may happen, but not as a typical feature of competitive behaviour. Firms rather tend to adjust prices following the dominant leader.[14-18]

Product differentiation and promotional efforts become the major focus of competitive dynamics. From the firms' point of view, business strategies centre on advertising practices that are most effective in securing high market shares and establish conditions for a lasting position of market power. The characteristics of infant-milk markets make certain promotional schemes particularly attractive because they directly create important captive markets for breast-milk substitutes. Maintaining high market share and market power position depends on the firm's ability and latitude to manipulate the breast-milk substitute's demand-creation parameters. Given the effectiveness of specific promotional and advertising methods to generate brand product consumption, marketing strategy enhances market growth and market power.

With a high market share and market power, the leading firms may obtain higher profits and be able to continue supporting high levels of promotional expenditures that effectively defend their market position when needed, or further expand their market development and demand-creation efforts. Implicit is the establishment of high barriers to entry that maintain a high degree of market concentration.[19-21] At the same time, large resources can be devoted on promoting an extended line of differentiated products.

Intensive marketing efforts lead to health professionals' acceptance of the firm's products, further consolidating market positions and the oligopolistic structure of the market.[21] Studies in the industry show that high levels of advertising and promotion are associated with above-average levels of profits, supporting the hypothesis that advertising and promotion are important sources of monopoly superprofits.[22] It pays to advertise. High concentration is associated with higher prices and above-normal profits.[14,23,24] In fact, effects of both concentration and advertising on profitability tend to reinforce each other. The more concentrated the industry, the higher the level of advertising and promotion and their effect on profit rates as shown in the US food-processing industry, as well as in cross-sectional studies of industry in developing countries.[25-27]

The high degree of national and world-wide market concentration for infant-milk products is a source of concern first from an ecomonic point of view. The issue with which health professionals must deal directly, however, is the effect of competitive pressures on adoption of artificial feeding and demise of breast feeding.

Infant-milk markets and barriers to entry. For most infant milks, which do not have specific therapeutic qualities and are closely similar in quality, the role of promotional efforts to increase sales becomes even greater. Differences between many formula products are of only minor medical significance, so that the inherent quality of infant formula products plays a relatively small role in the purchase/use decision as compared with prescription, and even some non-prescription, products.

The merits of a particular formula over its competitors consequently relies to a great extent on the firm's promotion, differentiation, and brand image activities. Market behaviour focuses on services for paediatricians and maternity hospitals, especially through the supply of free samples, so as to generate a rigid demand for the product, and to capture secure market shares. At the same time, these promotional activities erect very high barriers to entry, as shown below.

Initially high promotional expenditures are invested to get the infant-milk products accepted as widely as possible, and to obtain exclusive contracts with public health services. These unwittingly become commercial agents at the national level for the products at no cost to the firm. Competitive promotion now becomes unnecessary. Price competition also plays a minor part in competitive behaviour because price is not where substantial competitive advantage over other brand products can be gained. Rather, higher prices may be an additional selling argument to build the image of a 'highly scientific product'.[24] As promotional activities lead to highly concentrated markets, lack of price competiton in the industry is propitious to higher prices and bigger profits. The industry formula market structure evolves from business strategies that tend to create very high barriers to entry from the start. Once a large market share and market power are obtained continued dominance over the market becomes a self-sustaining process. Because infant formula is at its mature phase in the product life cycle,[28] most competition arises among 'me-too' products. Experience shows that, once established, the leading market position is well secured. Moderate increases in promotional expenditures can quite easily reinforce brand-name appeal and doctors' brand loyalty, preventing the 'me-too' competitor in succeeding without vigorous efforts to promote the product or to gain control of important distribution channels, such as the public health system.

Promotional competition constitutes a barrier to entry, especially because the infant-formula market is a brand type of market. To persuade practitioners to switch brands is difficult. Brand names form the basis for promotional activities and further reduce price competition pressures. Entry barriers against similar products can be maintained because of brand-name preferences are created in the target market groups, just as they are in the drug or cigarette markets.[28]

Brand names seem to indicate substantial differences in the composition of products, usually associated with some scientific endorsement. these differences are actually non-material, but use of esoteric names reinforces the subjective 'image' of specific infant formulas and prevents effective use of generic products (Table 12.5).

TABLE 12.5   *Brand names of breast-milk substitutes (Partial incomplete lists). Leading firms, 1980*

Breast-milk substitute company	Parent country	Parent company	Brand names
Ross Laboratories	USA	Abbott Lab	Similac, Isomil, Advance
Mead Johnson	USA	Bristol Myers	Enfamil, Prosobee, OLAC, Sobee
Wyeth Laboratories	USA	American Home Products	SMA, S26, Bonna
Gloria—S.A.	Peru, Mexico, Brazil, Philippines	Carnation	Gloria; Carnation
Foremost International	USA	Foremost McKesson	Campo Verde, Angel Dos Nenes, Dutch Glory, Milkman, SO–60
Borden	USA	—	Klim, Eagle Brand
Ursina-Frank	Federal Republic of Germany	Nestlé, S.A.	Guigoz, Ursa, Nativa, Nidal, Guigoz 1, Guigoz 2
Nestlé S.A.	Switzerland	Nestlé S.A.	Pelargon, Nido, Nan-Lactogen (Nidal), Nestogen, AL110 Nespray, Eledon, Prodietun
Allgauer Alpenmilch	Federal Republic of Germany	Nestlé, S.A.	Alete
Wander	Switzerland	Sandoz	Ovaltine, Elacto
Dumex	Denmark	Canada Development Corporation (CDC)	Dumex, Mamex
Farley's	UK	Glaxo	Ostermilk, Sunshine, Vitamilk, Golden Glaxo, Osterfeed, Farex
Cow & Gate	UK	Nutricia	Cow & Gate, Formula M, Cowlac MilkFood, Baby Milk Plus, Trufood

TABLE 12.5 *(cont.):*

Breast-milk substitute company	Parent country	Parent company	Brand names
C.C. Friesland (C.C.F.)	Netherlands	Frico Co.	Frisolac, Dutch Baby (Bebe Holandes), Frisocrem, Frisoland
Lijempf	Netherlands	Wessanen	Bebelac Z18, Z12, Z28, Sabelac, Frisania, Humanized N°1, Modified 18, Vita
Gallia	France	B.S.N. Gervais Danne	Gallia, Nursie
Diepal	France	B.S.N. Gervais Danne	Alma, Lacmil
Fany Oemolk	Austria	VARTA A.G.	Nono, Famny
Erba Carlo Nutr.	Italy	VARTA A-G	Dodolac, Auxolac
Milupa S.A.	Federal Republic of Germany	VARTA A-G	Aptamil
Nutricia S.A.	Netherlands	—	Almiron, Farilacid, Olvarit, Nutrilon,
France-Lait	France	France-Lait	France-Lait
Morinaga	Japan	Morinaga	Crown Dia—G
Meiji	Japan	Meiji	(Meiji) FM-U; FA; Soft Curd; Milk Formula; Modified Milk
Snow Brand	Japan	Snow Brand	Neomilk, P-E, P7A
Domo Bedum	Netherlands	Domo Bedum	Gitana, Gitanalac, Safety Milk
de Melkindustry Veghel* (DMV)	Netherlands	DMV	Little Man, Best Test, Elk
Holland Canned Milk Industrial	Netherlands	DMV	Alaska, Wadi Fatma, Coast, Friesan girl
Syntex	Panama/USA	Syntex	Mull Soy, Cho-Free
Leo de Winter	Netherlands	Coberco	My Milk, My Boy
Baker	UK		Ludu
NCZ	Netherlands	Coberco	Lita
Brabania n.v.	Belgium	Brabania	Syllac
Lidano	Denmark		Linolac; Humanized Baby Milk; Lidamin; Lidacid; Lidarina
United Dairy Men	Netherlands	CMC	Farm
Arinco	Denmark		Arinco

[1] Infant formula, including powdered and evaporated milk marketed as infant formula.
* Zuid Nederlandse Milk Industry (DMV): joint venture with Wessanen.

Levels of promotional expenditures and profitability. In general, promotional expenditures for infant formula can be high when the market conditions and associated business strategy require it. In developing countries, intensity of promotion may also increase when supplies are plentiful, and decrease when supplies are not available due to constraints in obtaining foreign exchange or cows' milk.[29] Although little direct information is available on specific promotional and marketing expenditures for an infant-formula product, informed observers of the industry estimate these costs to be no less than 10–15 per cent of sales value on the average. These estimates fall well within average companies' promotional efforts (Table 12.4). Because infant formula is a high-volume sales item, such percentage can be considered to be comparatively high and to represent unusual levels of expenditure for a food item.

When competition is aggressive, to keep market share in growing markets, or to increase it in mature markets, the promotional outlays may be even more inflated. Percentages as high as 30 per cent of sales value for selling and administrative expenses were attained in the 1950s and 1960s in US companies in the USA.[30] Considering that levels of free sample donations alone presently reach 7 per cent of total sales in some developing countries,[31] it seems that considerable marketing efforts continue to exist under certain market conditions. In Kenya, subsidiaries of infant-milk transnationals spent an average of 6 per cent of their sales revenues on mass-media advertising alone.[32,33]

With respect to profitability, data are even more shrouded in secrecy.[32-34] However, according to financial observers of the US formula industry, present margins of 20–25 per cent on the domestic market are the same as those obtained two decades ago.[30] Evidence suggests that profits from infant-milk products are not significantly different from those obtained from other pharmaceutical sales. In international operations, profit-margin estimates range from a low 20 per cent to a high 35 per cent.[34] The high-volume nature of infant-milk sales suggests that these products generate very profitable operations and are major contributors to overhead expenses. Information from some companies indicates that returns on investments in the formula business are similar or even higher than for the prescription pharmaceutical industry with annual operating profits reaching 40–50 per cent return on investments.[34]

Business expansion of the infant-formula industry

Growth objectives of the infant-milk industry are met through a series of possible strategies that vary according to the conditions prevailing in given product markets. The potential for increased sales basically depends on the ability to raise prices or to expand sales volume.

The first type of strategy is followed when sufficient market power is achieved, or follows from the strategy of the market leaders in the mature stage of the market development. The second type of strategy is adapted to a rapidly growing market, when number of births is increasing, or the rate of product penetration improving. The company can achieve its basic objectives simply by keeping a sizeable market share.

In mature markets, where overall market volume remains relatively stagnant, a company's sales growth can be achieved by increasing its brand product share of the market. The basic way to meet such goals is to step up promotional competitive efforts to carve out a larger market share. Another alternative is to acquire greater control of the market through the merger or acquisition of other infant-milk producers.

When the company's control of the market is already substantial, an alternative strategy consists of trying ot increase demand for the company's products. This can be done by multiplying the product's line, introducing differentiated brand products on the market, or simply trying to extend the concept of product usage among consumers. Another option to maintain sales growth can be sought by entering new markets abroad which offer high sales potential.

All these scenarios, summarized in Table 12.6, illustrate possible ways by which an infant-milk processor may try to maintain satisfactory growth of its sales of breast-milk substitutes. Each of these strategies need to be evaluated in terms of relative profitability for the company and respective chances of success. Depending upon the product's life cycle and market develpmments in its major traditional market, a company will have different incentives to put emphasis on one or a combination of these alternative strategies to pursue its corporate objectives.

TABLE 12.6 *Expansion strategies in the infant milk industry: framework*

Type of market	Strategic objective	Business operations (examples)
Developed country—'home' market	Increase market share	Intensify promotional competition: maternity services and product development (disposable bottles, nipples), educational promotion, gifts, give-aways, 'starters'
	Use market power	Raise prices
	Demand creation	Product differentiation: new brand, packaging, '2nd age', mineral/vitamins added Form of presentation: powder, concentrate, ready-feed, extend concept of product use one year, 'specialty' milks (multiplication of brands and appeals)
Developed country—'New mature' market	Market entry	Acquisition/merger, 'specialty' milks
Developing country	Entry/demand creation	Imports/repackaging/production, Government contract
	Increase market penetration/market share	Promotion, public; promotional competition, medical

With this framework in mind, the development of business strategies of the major infant-milk companies can be examined. The analysis of the market conditions for infant formula in developed countries is especially important because traditionally they constituted the major sales outlets for the companies, and hence their main source of growth. The development of a situation where continued 'home' expansion becomes increasingly difficult means that there are greater pressures for the companies to seek expansion abroad. With such unfavourable conditions apparently emerging throughout all major developed countries, 'expansion abroad' strategy then means increased efforts to expand sales in developing countries where the market penetration of breast-milk substitutes is still incomplete from a business point of view.[11]

Recent developments in the infant-milk market structure.
Modified cows' milk product destined to replace mothers' milk have been commercialized for a long time. Most brand names of major infant formulas currently on the market were already established by the late 1930s.[34]

The post-war baby boom, combined with the progressive abandonment of breast-feeding practices and increase in artificial feeding with evaporated milk, caused a major market expansion in the post-war period.[35] Improvements were made in product quality and presentation of the breast-milk substitutes from powder to liquid concentrate, and finally 'ready-to-feed' solutions. Currently in terms of product life cycle, the infant-milk products in the developed countries seem to have reached their 'mature', declining stage.

At the time market penetration of formula in North America and Europe was being completed, a fundamental change occurred in the market conditions. With the sharp decline in the fertility rate, the product market's basic support, the number of newborns, shrunk considerably. The effects of this change were first felt in the USA during the late 1960s, then Europe, and lately in Japan (Table 12.7).

TABLE 12.7   *Live births in selected developed countries, 1960–78*

Year	Thousand births						
	Italy	France	UK	GFR	NL	USA	Japan
1960	910	816	—	968	239	4157	1606
1965	990	862	—	968	245	3700	1823
1970	901	847	920	980	237	3730	1934
1971	906	878	901	778	227	3555	2000
1972	888	875	884	701	214	3258	2030
1973	875	799	799	635	195	3137	2092
1974	869	801	737	626	186	3159	2030
1975	828	745	697	600	178	3144	1901
1976	784	720	675	602	177	3167	1832
1977	742	744	656	583	173	3312	1755

GFR = German Federal Republic.
Source: *Demographic Yearbook, Special Issue*. United Nations, New York (1979).

For instance, the number of births in the USA dropped from a peak 4.3 million in the early 1960s to a mere 3.1 million through the 1970s, after 1960 forecasters had predicted 7.0 million births for 1975. In Europe, the number of births fell by 20–60 per cent between the late 1960s and late 1970s. Under such conditions, the potential for growth in the volume of sales became limited (Table 12.8). In addition, the proportion of breast-feeding mothers has increased significantly in most developed countries since the mid-1970s.[4,36] To maintain previous rate of sales expansion, companies were compelled to adopt aggressive non-price competitive marketing strategies.

TABLE 12.8   *Impact of the decline in birth rate on formula business. Selected companies*

Country	Company	Annual report, year	Comments
USA	Mead Johnson	1972	Enfamil was somewhat affected by the declining birth rate, although business (in dollars) continues to grow
France	BSN-Gervais	1975	Unfavourable outlook for baby foods, characterized by a reduction in the number of births
Germany	Varta A.G.	1976	The stagnating birth rate in Germany inhibited the natural growth of Milupa products in 1976. Continued growth of Milupa is being inhibited by declining birth rates also in countries like France, UK, Italy, and also large parts of Eastern Europe
UK	Unigate	1976	Because the birth rate has been falling in the UK sales have fallen too but we expect people will want more infant milk food in overseas countries. For this reason, we made Cow & Gate part of the international division
UK	Glaxo	1976	In the UK, the decline in birth rate continued and the total market for infant milk foods diminished. However, the share held in total by Ostermilk improved and Farley Health Co. had a profitable year
		1977	Although the UK market continues to shrink slowly, as birth rate declines, Farley had a profitable year
		1978	According to the firm, the UK market is declining because of lower birth rate—but Glaxo is well established in some developing countries (Dafsa)

Source: Compiled from corporate annual reports.

Intensification of promotional efforts. In the 'mature' markets of industrialized countries, the development of market conditions has seen an intensification of promotion and increase in market concentration. For the companies that cannot directly use their market power to increase sales, by means of price increase for instance, the first and immediate response to slower market growth has been to step up promotional activities.

In this respect, the USA and European experiences confirm the key role played by hospitals and paediatricians in gaining a larger market share. A suit-action brought in 1975 by an American producer against its competitors claimed their marketing practices included:[37]

1 The supply of infant formulas to maternity wards at predatory prices designed to eliminate competition, or at no price whatever to the hospitals.
2 The supply of nipples, nursers, and other products related to the use of infant formulas in maternity wards, at predatory prices designed to eliminate competition, or at no cost whatever to the hospitals.
3 The payment of cash grants to maternity hospitals in addition to the above predatory prices and give-aways.
4 Subsidizing such predatory prices and give-aways to maternity hospitals with profit from the sales of infant formulas at substantially higher prices through chains of distribution after discharge from maternity hospitals.

Similarly, as reported by a chief sales executive of a leading infant-milk company:

In France the infant-milk market is saturated. There is a ferocious fight at the level of the maternity clinics because that is where the formula-feeding process starts. You know, one tries to take over the strongholds. They (maternity units) are the Dien-Bien-Phu (of the infant formula market).[38]

The emphasis on the maternity hospitals as a focus of promotional activities has led to higher levels of sophistication in the approach and methods used. A wide range of promotional services are provided to these institutions. Besides the usual gifts of equipment or product samples, the preferences for the company's brand products is sought through elaborate education or assistance–promotion programmes, such as 'Hospital Formula Feeding Systems' (Table 12.9). Research and product innovation in the industry itself focused on items such as disposable bottles, universal nipples, and other equipment that facilitate the ward routines for formula feeding, and thereby the introduction of the company's products to the mothers. Systematic approaches to 'personal selling to maternity hospitals' have been developed to 'gain' all major key-deciders, including nurses. As explained in a major infant-formula-company training manual for salespersons:[39]

If they are sold and serviced properly, they can be strong allies. A nurse who supports Ross Lab. is like an extra salesperson. She sees the doctor every day and can influence his/her choice of formula. There is more than one story of a nurse being the key to our getting new hospital business. . . . Get to know all nurses. Occasionally visit the 3 to 11 and 11 to 7 shifts. The gals on these shifts will appreciate the attention.

TABLE 12.9 *Breast-milk substitutes: corporate promotion services to maternity hospitals. Ross Hospital Formula System (excerpts)*

Features	Benefits
The discharge packages	
Ross provides a selection of nine discharge packages.	Provides flexibility for the physician to provide the kind of feeding programme he prefers.
Discharge packages are wrapped in clear plastic.	This makes it easier for nurses to point out and explain contents without having to open package.
Complete instructional literature is provided.	This helps in educating mothers about feeding procedures and about the importance of good nutrition for her baby.
The Services	
Ross provides a *disposable newborn measuring tape*.	This is a simple, clean, and convenient way to take care of this need in the nursery.
Ross provides a complete catalogue of feeding products and services.	Hospitals may have a quick reference source to know what's available from Ross.
Complete series of *teaching aids* available.	Nursing classes may have useful slides and pictures for educational requirements.
Ross provides many nursing inservice aids.	These provide a quick, easy-to-use source of information for nurses in the nursery and delivery room.
Ross publishes several *newsletters*. Topics covered are paediatrics, family practice, obstetrics, hospital administration, feedings, public health and nursing.	These provide hospital personnel with the news of latest developments in the various fields. They are sent free of charge to those requesting them.
Ross provides a series of informational booklets which cover many aspects of child-raising.	Hospital personnel have access to a source which can help answer many of mothers' questions.
Special printing available on request.	Done by the Ross Printing Department, these can help hospitals with printing problems unique to a particular hospital.
Ross maintains *a hospital planning department staffed* by experienced planning experts.	These services are available free of charge to hospitals contemplating additions to or modifications of their facilities.
Name pins and desk name-plates provided free of charge.	Hospitals may obtain a necessary item easily and quickly.
Similac samples are made available to physicians and ancillary medical personnel under guidelines of POF policy.	*Doctors get Similac products at no charge*. This provides a means by which doctors may sample Ross products with both his patients and his own family.
Emergency supplies of hospital products are maintained by each Ross Territory Manager.	Should emergency situations occur at the warehouse, hospitals may call on their Ross Representatives for immediate assistance, whether day or night.
Ross Territory Managers can readily obtain any product information.	Special questions that may come up can be handled quickly.

TABLE 12.9   *(cont.):*

Features	Benefits
Ross Territory Managers make regular calls on hospitals and physicians.	Hospitals and physicians have ready and regular access to the person who can help with problems or questions.
Ross Territory Managers work closely with hospital purchasing agents.	This helps prevent stock outs.
Ross provides initial and continual training of hospital personnel in the use of the Ross System.	This helps personnel remain aware of new developments and assures that new personnel are quickly informed about the Ross System.

Company detailpersons are trained to subvert maternity ward rules when not propitious to the company's interests. They are taught, for instance, how to get direct access to the premises and mothers when not allowed to. Although the leading US infant-formula company is reputed to be among the most restrained in its marketing practices, its hospital sales training manual and sessions instruct explicitly how 'to sell when you are not allowed on the floor' (Table 12.10).

TABLE 12.10   *Breast-milk substitutes: personal selling to maternity hospitals. Excerpts from training manual used in salesmen training, Summer 1978, USA*

Selling hospitals when you're not allowed on the floor

Show your catalogue of products and services to P.A.'s or whoever you routinely contact. Tell them the nurses may be interested in something and ask permission to see them.

Call the nurses and ask them to meet you in the coffee shop and discuss RHFS and services with them.

Mail literature and service items, etc., to nurses with a B.R.C. to you.

Never miss the chance to get permission to demonstrate a new product or service.

Get a physician interested in your products and have him make a special request that you be allowed in the hospital.

Find that one interested person who can open the door for you.

Find *any* excuse to get up on the floor. For example, you can check booklets, mail list, formula, etc.

Don't miss the chance to do something extra for the hospital to show you are truly interested in its well-being.

Leave your card with P.A.; you never know when they may need something in a hurry and there's always the chance competitive salesman may be unavailable.

If you can't get up in the nursery find other routes into the hospital such as Administration, Director of Nursing, In-Service Director, Pre-Natal Class Instructor, etc. We have a unique product or service for everybody.

Call on the hospital in the evening and see the night shift (this is risky but effective).

Neutralizing or winning over that one stumbling block

Sometimes one person stands in your way of getting the business. When this happens you must identify (by proper questioning) and neutralize whoever it is.

Source: Reference 39.

Actions geared towards expanding consumption through multiplication and differentiation of products have also been increasingly adopted. First, 'second age' products such as Advance or Nestogen were proposed as a logical follow-up of the formula used for the first months of baby life. Companies stressed the nutritional value of breast milk and the importance of good nutrition to put a higher value on the artificial products supposedly 'closest' to mother's milk.[40] This argument serves to promote the use breast-milk substitutes as long as possible, at least until 12 months of age, e.g. Similac, Enfamil.

The presence of certain minerals or vitamins, or even honey, is also used as a sales argument to differentiate the products. But perhaps the most disturbing development in the differentiation drive is the addition of specialty infant milks, such a hypoallergenic formulas. These are usually introduced following the infant's intolerance to normal 'well-baby' formulas. In these cases, artificial feeding directly caused the pathological conditions, where hypoallergenic products are needed. In developed countries, sales growth of these products has been notable in recent years. They accounted for 20 per cent of breast-milk substitutes sales in the USA—US$100 million in 1979—compared with only 2 per cent a decade earlier.[41]

Mergers and acquisitions. Stiff competition in the slowly expanding 'mature' markets has resulted in a significant transformation of the industry structure through numerous acquisitions within and across borders. In the USA and Europe, the smaller independent producers disappeared, either driven out of business or absorbed by the larger formula companies. In the early 1970s, two national producers and most regional producers of infant formulas were eliminated in the USA, leaving the market under the control of the three largest companies.

The direct consolidation of sales between two companies is another way to increase market share and power. Mergers of milk companies occurred particularly in Europe (Table 12.11). This business strategy had an effect not only in a national market structure, but also internationally, such as with the Nestlé–Ursina Franck merger, that augmented market concentration in various markets at the same time.

In fact, firms' entrance in 'mature' foreign markets appears extremely difficult without outright acquisition, with the possible exception of specialty milks. The competitive advantage associated with a long-standing relationship with health professionals has proven to be an insurmountable barrier for entry in such 'new' markets. Major entry is done through acquisition of local dominant firms. Even Nestlé, the leading world producer, failed in its efforts to enter directly the USA, German and French markets, and therefore proceeded through acquisition. This may be a major reason behind the intensity of promotional efforts observed in developing countries, even when the current local market is small. A secured market position tends to be long lasting, once established.

Leading formula producers themselves were acquired or merged with larger diversified, rapidly growing, internationally oriented firms. This

TABLE 12.11 *Acquisitions and mergers in the breast-milk substitutes industry. Leading transnational corporations*

Parent company	Country of origin	Acquisition merged/ subsidiary company	Country of subsidiary	Year merger/ acquisition
Borden	USA	Morrell Soule	USA	1927
American Home Products	USA	S.M.A.	USA	1929
Nutricia	Netherlands	Cow & Gate	UK	1980
Unigate	UK	Cow & Gate	UK	1959
Canadian Development Corporation	Canada	Dumex		1977
Eastern Asiatic/ McKerson	USA Denmark	Dumex International	Denmark	1964
Abbott Laboratories	USA	M & R Laboratories	USA	1964
Bristol Myers	USA	Mead Johnson	USA	1967
Mead Johnson	USA	Six acquisitions abroad	USA	1960s
Glaxo	UK	Farley Health) Foods	UK	1969
Beecham	UK	Horlicks	UK	1969
Sandoz	Switzerland	Wander	Switzerland	1970
Domo-Bedum	Netherlands	Funo, Mayo & Llemff plant	Netherlands	1970
B.S.N.–Gervais	France	Evian S.A. (Diepal, Gallia)	France	1970
Montedison	Italy	Carlo Erba	Italy	1971
Frico	Netherlands	Coop. Condensfabriek Friesland C.C.F.	Netherlands	1971
Nestlé S.A.	Switzerland	Findus	Sweden	1969
Nestlé S.A.	Switzerland	Ursina-Franck (Allgauer Alpenmilch- guigoz)	Federal Republic of Germany	1971
Nestlé S.A.	Switzerland	Beechnut Products	USA	1979
Wassenen	Netherlands	Lijempf	Netherlands	1973
Varta (Busch-Jaeger)	Federal Republic of Germany	Milupa S.A.	Federal Republic of Germany	late 1960s
Varta	Federal Republic of Germany	Erba Carlo Nutritional (?)	Italy	1975
Varta	Federal Republic of Germany	Fany Oemilck (?)	Austria	1975
Varta	Federal Republic of Germany	(?)	France	1975
Syntex	USA	Borden Pharma- ceuticals	USA	1975

Compiled from corporate annual reports and trade sources.

TABLE 12.12 *Recent trends in overseas operations by major infant-formula companies*

Transnational company*	Foreign sales, (percentage total sales)					
	1950	1965	1970	1975	1978	1980
Mead Johnson (Bristol–Myers)	14	—	—	—	48	—
Ross Laboratories (Abbott)	—	3	16	20	—	25
Milupa S.A. (Varta)	—	—	4	—	34	—
Nutricia	—	—	—	32	40	75

Notes from Annual Reports

1969	Abbott (Ross Laboratories)	In international operations, Similac showed greatest growth, with sales increasing even more rapidly than Similac sales domestically
1971	Abbott (Ross Laboratories)	Overseas sales volume of Similac increased more than 44 per cent, aided by extension of the product line into Australia and 10 African countries. Vigorous sales expansion occurred in two older markets, the Middle East and Canada
1973	Ross Laboratories	In the past year, Similac sales grew at their fastest rate in Africa and the Middle East
1974	Ross Laboratories	Sales of infant formula products more than doubled in 1973 in the Middle East and Africa
1979	Abbott (Ross Laboratories)	Over the past years, international sales have grown at an average compound rate of 17 per cent
1972	Glaxo	Significant improvements in sales and profits in Middle East
1977	Glaxo	Sales in Asia and Latin America doubled between 1973–74 and 1976–77
1976	Unigate	Exports reached highest levels ever in 1976. Cow & Gate brand in baby foods has benefited well from good planning and our wide contacts with the medical profession
1978	Unigate	Significant progress made by Cow & Gate export section
1978	Snow Brand	Snow Brand Powdered Baby Formula is increasingly recommended by doctors and nurses abroad and has received wide consumer acceptance
1976	Borden	Whole milk powder operations in Ireland and Denmark had outstanding results as world-wide markets for Klim brand powder continued to show substantial growth
1977	Borden	Klim, mainly European based but entirely an export operation to developing countries, enjoyed best year ever, with sales setting new records
1979	Bristol Myers	Excellent gains world-wide in nutritional produce sales were reported in 1979 and represented the most significant impact in sales and operational profit growth of the segment

* Subsidiary of Division responsible for infant nutritional products.
Source: Annual Reports.

transformation, particularly active in the late 1960s in the USA and up to the late 1970s in Europe, reinforced the relationship between the formula industry and world-oriented transnationals. This close association has direct implications for the marketing of infant formulas in developing countries. When faced with declining possibilities for rapid expansion in their traditional outlets, growth abroad becomes a logical and financially feasible business response which would not have been an option for smaller, local firms. A close examination of the behaviour of the major world infant-milk companies confirms that such a response has become increasingly important in recent years, even when 'home' market sales exhibited reasonable growth (Table 12.12). The establishment of separate international divisions within the major breast-milk substitute producing firms in the late 1960s and 1970s reflects the growing importance of overseas operations.[40]

Competitive pressures to expand market development and sales growth in developing countries now are reportedly increasing. For some of the leading infant-milk producers, developing countries' sales account for as much as 75 per cent of total sales. The Third World is seen as the major growth market, with firms stepping up their marketing efforts to gain a market position in the boom forecast there for the 1980s.[42,43] Meanwhile, the hotly competitive baby-food companies are under pressure from the recession, and from stagnating markets in developed countries. The baby-food market in the USA, for instance, is expected to decline 4 per cent by 1990.[44] Understandably, leading corporations have been particularly concerned about the negotiation of a code for ethical practices in the marketing of breast-milk substitutes.[43,44]

Communication strategies for infant-formula marketing

Determining factors in the adoption of bottle feeding and formula consumption

The position of the infant-food industry in the crucial choice of infant feeding has become entrenched through a number of promotional strategies. Competitive pressures induce leading producers to augment their demand-creation activities for breast-milk substitutes and especially to resist any limitation put on them. A detailed analysis of the dynamics of marketing behaviour in the industry is presented here to allow careful assessment of the forces in play.

As in many other industries, the behaviour of the infant-milk industry is derived from the influence it can have on the demand for the product. The adoption of formula-feeding practices depends on numerous factors, such as the characteristics of the food, the nature of sales outlets, the socioeconomic status of the buyer, and his or her motivations. However, the breast-milk substitutes market has some unique characteristics for a food product that have a profound influence on the industry's marketing strategies.

Producers play an active role in creating the demand for their brand formula products, and so invalidate the belief that the free play of demand and supply determines the optimal appropriate allocation of resources. The concept of demand is not an appropriate tool to analyse this market. The identification of major elements that contribute to introduction and adoption of the formula products constitute a better basis for analysis. By and large, it can be said that the infant-milk companies have intervened in the natural relationship between a mother and her child in such a way as to modify it by introducing their product. As a result, the consumption level of the formula products (or use of bottle feeding) is strongly influenced by companies' marketing activities. The basic mechanism by which companies succeed in generating demand for their formula products is by their action through the medical profession.

Because most infant formula is introduced after a medical recommendation, usually a 'prescription' at the release from the hospital/clinic where birth took place, the product has achieved the status of a quasi-prescription drug. By transforming what is merely a product designed as a food—and not designed to correct or palliate existing medical problems in most cases—into a quasi-prescription product, the characteristics of consumption have become considerably different from ordinary food consumption goods.

Four major determinants in the introduction and adoption of bottle and formula feeding can be distinguished, relative to the special status of formulas as 'quasi-prescription drugs'.

Dependency effect. The introduction of the bottle and formula in the newborn feeding leads to a rigid pattern of consumption. After its initial use in the hospital feeding routine, infant formula is perceived by the mother as a necessity for the good health and nutrition of her child. Further, when she receives the 'prescription' or medical recommendation for a given formula product, she is led to see its observance as necessary to her baby's health. As such, she has been put under a particular obligation to buy the product. At the same time, the whole breast-feeding process has been greatly undermined, creating an irreversible dependency on breast-milk substitutes.[36,46-50]

Directed adoption of formula feeding. Adoption of formula feeding may be depicted as a 'directed' consumption pattern, where the principal decision maker is usually not the consumer, but the paediatrician, and indirectly the industry. The introduction of the product and the subsequent purchase decision rest mainly with the authority of the medical personnel. Because medical personnel and in particular paediatricians and nurses are looked upon as providing necessary guidance in baby care, they are perceived as having unique knowledge of the properties of the products they recommend. They are in a position to impose forcefully on mothers what is 'right' to do and the mother's full confidence in the particular brand is so well established that high loyalty to the recommended brand is maintained in future purchasing decisions.[51]

The medical personnel's influence on a mother's decision is further reinforced by the highly 'professional' setting of maternity hospital wards, where a particular brand of infant milk is usually introduced to her. The environment itself conveys a maximum and lasting effect to the product image and its perceived benefits for the mother and baby.[51]

Consumption–creation effect. The consumption of infant formula is a self-generating process. Introduction to infant formula and the bottle in the early days of life inhibits the successful establishment of a normal breast feeding. Supplementation generates less breast-milk consumption by the baby and reduces its normal supply, forcing the mother to use more of the 'best' substitute for mothers' milk. It also jeopardizes mothers' confidence in their ability to breast feed their babies, creating the psychological attitudes that result in drying up of the milk and early weaning from the breast.[46]

Price-insensitive decision and necessity effect.[24] Introduction of infant formula is a price-insensitive decision. The recommendations for a particular (brand) product are usually made in isolation from cost considerations for three reasons. First, the economic burden associated with the medical choice tends to be thought irrelevant to good practice. Because health is affected, quality and availability of the product have more importance than price. Second, the main 'deciders', paediatricians or maternity staff, do not bear the costs implied by their decision. A third element in price insensitivity is the fact that for infant milk, consumption at the time of initial recommendation is much less than the prolonged consumption requirements it generates in the months ahead.

Furthermore, the consumers who pay for the product and bear the costs are not in a position to evaluate the cost implications of the decision. This is due to the dependency effect, the high credibility placed in the doctor's decision, as well as in the 'free samples' or 'starters' they may receive initially.

Promotion strategies for infant formula

The above determining factors in the adoption of infant formula are of fundamental importance for the marketing strategies of the industry. They further reinforce a tendency for lack of direct price competition among companies and provide all incentives to consider infant milks as products similar to ethical drugs, accentuating the role of promotion in their business strategy.

Existence of high and stable levels of concentration in the industry where branded products dominate suggests that such promotional activities permit the establishment of market power positions.[13]

Given the quasi-prescription nature of the infant-milk market and its demand-creation characteristics, an escalation process in promotional activities between leading companies is generated. The intent is both to introduce their brand formulas as early as possible to each mother, and second to establish a dominant market share early in the product development cycle, and thereby secure a position of long-term market power. The influence of such practices on health services and the adverse impact on breast feeding is often considerable.

This overwhelming role of certain specific promotion efforts in generating infant-milk consumption calls for a detailed examination of how the companies' communication strategies to sell their breast-milk substitute products are elaborated and how they work. Appropriate policy measures concerning regulations of industry practices can be then assured.

The design of communication business strategies

Marketing strategies are designed to increase the volumes of sales and improve the market share of the company's products. Promotional and marketing techniques are conceived not only to sell a specific product but also—as in the formula market—to create additional demand for it. Central to these efforts is the communication process that is used to stimulate more sales.[52] By definition the communication mix is constituted by the series of activities that attempt to convey messages triggering more purchasing of the product. Their selection is to influence target groups' decisions, and depends on a careful analysis of the market conditions. In this process three questions are of central importance:

1 To whom shall the communication messages be transmitted; who is the principal target group?
2 How may these messages best be transmitted; what are the appropriate channels of communication?
3 What types of messages will be efficient to achieve the marketing objectives; that is, increased general consumption of the branded products?

Specific answers to these questions depend on the product characteristics the organization of the firm, and the details of the buyer–decider-influence patterns, to mention a few relevant factors. Thus, it is clear that there are not one but several marketing strategies that emerge as all these factors are considered in each specific market.

Definition of target markets

Paediatricians and maternity hospital staff are central market targets. In the pharmaceutical sector, the markets addressed in priority by the communication strategies are the physicians because of their pivotal role in product purchase decision. Because of very similar characteristics for determining infant formula consumption, *the same emphasis* characterizes the marketing behaviour of the formula companies.

The central role of paediatricians, obstetricians, midwives, and maternity hospital personnel in product use gives them a special importance in the companies' promotional activities. As discussed earlier, not only do they prescribe the product, but in a sense create its future consumption.

In the drug market the disease prevalence and size of the population greatly determines the level of consumption. For infant formula, however, the industry has the ability to influence consumption directly through the medical staff. Contrary to other conditions, where illnesses are outside the company's control, the incidence of infants that will need artificial feeding can be considered to be within the sphere of corporate influence. This is because of the lasting effect of early formula supplementation and the strength of paediatricians' prescription/endorsement on artificial infant-feeding adoption and product-brand choice. For this reason, the industry concentrates its promotion efforts on the target groups of hospital 'newborn' personnel and paediatricians.

This general focus may, however, be influenced, or rather complemented, by another consideration: the degree to which infant formula has achieved the status of a quasi-prescription among the population of mothers, and whether practices of prolonged breast feeding are still common in the population. In situations where a number of mothers are not fully integrated into the formal infant health-care system, communication strategies may also put emphasis on efforts directly geared towards these mothers through mass media and other direct public promotion channels.

Mothers as a target market have a special importance. Each product goes through different stages in its diffusion to the market, depending on the levels of market development for the product. Starting with its introduction, the market grows rapidly until it reaches general acceptance and maturity. When the market penetration is completed, the product has reached saturation of the market and begins its decline phase, as new products or new concepts are introduced.

In this framework, infant formula can be considered a mature product for the medical profession throughout the world in that it has become an integral and routine part of neonatal care and the prescribing behaviour of paediatricians/nurses to mothers. On the other hand, diffusion of the artificial feeding concept among mothers may still vary enormously. The predominance of medical influence is not uniform across nations, or even across socio-economic groups within nations. In many developing countries where the tradition of breast feeding is still strong, many mothers have had limited contact with the medical establishment, say, in rural areas, or even in urban areas among low-income groups that have little access to health-care facilities and no social security benefits. Many mothers in developing countries have been introduced to bottle feeding, but direct influence of medical personnel may still be limited. These groups of mothers, however, constitute a potential new market, where product penetration can be increased and sales augmented.

Applying the concept of product life cycle to breast-milk substitutes suggests that different target markets are chosen as focus of marketing strategies depending on the socioeconomic environment. Channels chosen for the communication strategy vary according to the nature of each target market. In developed countries, the emphasis is on promotion efforts and channels aimed at hospitals and doctors. In developing countries, marketing efforts directed towards the public are also utilized.

'Personal selling': major channel for 'selling' communications

To carry out their communication strategy, the infant formula companies may use four types of transmission channels:[52]

1 Advertising; the use of commercial media for non-personal presentation of the products.
2 Publicity; the stimulation of media coverage of significant news items about the company and its products, without paying for such presentations.
3 Sales promotion, or 'below-the-line' activities, that do not use media. These activities include shows, conferences, exhibits, gifts and free sample distribution, cash payments, and give-aways.
4 Personal selling, or the 'rifle' approach, with a direct oral presentation to the prospective buyer or influential deciders, including demonstrations.[53]

The selection of the most appropriate channel to convey the company's messages depends on a number of factors, such as the legislative regulations, the competitive situation, and the socio-cultural environment. The nature of the target market groups greatly influences such decisions. A major part of promotional efforts for infant milks as 'quasi-prescription products' are put into direct professional selling activities:

Few doctors have time to read professional journals: therefore other ways have to be found for pharmaceutical manufacturers to make their impact: medical detailmen, or direct mailing. Because of the personal contacts they maintain, representatives are the most important link between manufacturers and doctors. The emphasis is not upon the quick persuasion of doctors to prescribe the firm's products, but to indicate their quality in the process of building up goodwill over a number of visits.[54]

The sales representatives, or detailpersons, have been shown to be one of the most efficient means of influencing the prescribing behaviour, especially in obtaining the important 'repeat prescribing pattern'.[55] In LDCs, where diffusion of scientific journals is often very modest, detailing is likely to have even more convincing and powerful effects.

Detailing presents many advantages over advertising because of the way it works, such as being a two-way flexible communication that can be tailored to fit not only the individual paediatrician, but also the particular situation in which the interview is conducted, and can also respond directly to potential objections or negative feelings. Table 12.13 illustrates similarities and

TABLE 12.13   *Advertising and detailing: similarities and contrasts*

Characteristics	Advertising differences	Similarities	Detailing differences
Functional	One communication. Abundant 'noise', in the communication channel. Relatively inflexible. Good control over the message. Almost impossible for physician to avoid *some* exposure to the message	Both must be: understandable, interesting, believable, persuasive	Two-way communication. Some control over 'noise'. Can be tailored to the situation. Difficult to maintain company control of the message. Physician may refuse to see
Perceptual	Difficult to reinforce the message during the course of presentation	Both must penetrate sensory mechanisms of the physician, with careful selection of stimuli necessary	May stimulate all five senses as vary them selectively. May reinforce and repeat in a single call
Cognitive	Works primarily by *suggestion*. Primarily an *interest-arousing* technique	Both attempt to present firm and product as *different* and *better* than competition	May carry physician through *reasoning* process. May be a *problem-solving* technique
Feeling state	Single message may elicit varying feelings. No possibility to adapt feelings	Both attempt to induce favourable feelings	May evaluate and take advantage of either favourable or unfavourable feelings
Transactional	Primarily pretransactional, with post-transactional activity primarily limited to dissonance reduction	Both important as reminders to continue use	May also effect a prescription as a direct result of sales call. Possible to supply sample for patient in the office at the time

Source: Adapted from Reference 55.

contrasts between advertising and detailing in terms of the power of their selling messages.

The importance of these detail personnel in formula marketing is attested by their large number. For instance, a major US formula company had in 1961 a staff of 165 detailpersons, or about one per 100 paediatricians in the country.[56] In the Philippines alone, a study of the five major formula companies indicated that they employed in 1975 an average of thirty-five 'detailpersons'/nurses, enough to keep a strong pressure on professionals and also conduct extensive 'mothercraft selling' activities.[31]

The major functions that detailpersons perform are:

1 To sell the products by persuading the doctors concerning new products.
2 Further developments in use of existing products.
3 To distribute samples and product literature to doctors.
4 To tell of the experience of one doctor to another, an important message in the doctor's acceptance of the product.[55]

In the determination of the detailperson's salary, the company usually takes into account the evolution of the branch sales in their assigned sector, either directly (percentage or commissions) or indirectly (premiums).

Because the target market of paediatricians is clearly defined, identifiable, and relatively compact, it provides excellent conditions for 'direct selling' types of promotional efforts.[53] The group of paediatricians is usually segmented by the companies into different categories depending on their prescribing sales potential, for example, the number of infants they deal with:

The industry has been fairly successful in identifying the "high prescribers" market. These physicians are obviously in for different and more intensive promotion efforts than the average.[55]

Three 'categories' of health personnel were so identified by two of the dominant firms in the Philippines, depending on their potential for promoting breast-milk substitutes.[40] Various degrees of material incentives and treatments are tailored to each category. The amount of attention spent on each doctor depends on the magnitude of his/her prescribing potential. Beyond the traditional offer of free samples, sales efforts are complemented by cash grants, equipment, and sponsorships to attend conventions, and have included grants and contributions of a percentage of breast-milk substitute sales to professional associations.[57-59]

Other 'personal selling' activities. The first priority of the infant-formula companies is to influence the public and private maternity hospitals (Table 12.14). Maternity wards constitute, it cannot be overstressed, the *critical* focus of the infant-formula companies' marketing strategies. Although maternity ward purchases of formula represent a relatively small share of the total market, they play an essential role in market control. The maternity ward is where formula introductions and brand preference are most strongly developed. This is because of the influence of the birth and first feeding experience on the young mothers, and also because maternity hospital staff professionals are usually opinion leaders for mothers. They are 'top 1 priority' targets of the infant-milk industry. Recognizing this, companies have developed a whole series of promotional and management systems for formula rooms that aim at influencing hospital staff preferences by catering to some of their needs.

After routine supplementary formula feeding, the mothers leave the wards with the inevitable prescription that builds strong loyalty, desire, and need to use the breast-milk substitute brand product. In the USA, for instance, it is estimated that about 95 per cent of the mothers continue to feed their babies with the brand used in the hospitals, after their release. Early 'supplementation' seriously undermines normal lactation and breast-feeding process. When 'starter' formula cans are made freely available to mothers at the release from the ward, powerful additional incentives for bottle feeding are given to the mothers.

TABLE 12.14   *Role of maternity wards in the marketing strategies of the infant-formula industry*

Baby Formula is a high-volume item that is practically pre-sold to new mothers via physician/ hospital endorsement. *American Drug*, August (1970)

There is a relative scarcity in coop-money available to drug stores. The reasoning by producers is that 'Mothers are heavily dependent upon a doctor's recommendation when they buy formula for their children. It is almost like a prescription where one particular brand named by the doctor will be the brand the patient gets. So consumer advertising doesn't benefit formula manufacturers as much as does 'ethical promoting and detailing'. *American Drug*, April (1977)

Milk for infants made satisfactory progress despite falling birth rates. Some factors are more favourable, such as the increasing buying power in the LDCs and the rising number of births in maternity hospitals, where it is easier to reach mothers. This is due to the fact that the medical profession staff there is more likely to influence mothers with regard to the food most suitable for their babies. Nestlé, *Annual report* (1971)

Selling to maternity hospitals with services. If properly used, they can by *dynamite*. It is up to us to see we get the most out of it. Abbott Laboratories. *Detailmen hospital sales training manual*. July (1975)

We sell to a very broad market, especially infants' food. That gets wider and wider and wider—due to the clinics where mothers go to—that is where they are introduced to these products. Company executive, Nairobi, Kenya, 1973, in Senate Hearings (1978)

Mead Johnson is frequently introduced to *new* parents by their doctors or in the hospital with the Enfamil Discharge-Pak. Mead Johnson, *Annual Report* (1973)

The Similac Hospital Formula System is in use in more than 2500 maternity hospitals throughout the United States, where more than half of all US births take place. Abbott Laboratories, *Annual Report* (1968)

One way to control formula use in maternity wards is to obtain a contract with the public health system, or with social welfare organizations and give de facto control to a particular company of a share of the breast-milk substitute formula market. Such a contract provides a unique competitive position, and also it reduces the amount of samples that must be donated at industry expense since the public health authority usually pays, even if only at a discount, for the samples usually provided free. Wide diffusion of the products throughout the country is obtained without engaging in the difficult task of winning the preferences of local personnel in charge. The competitive advantages gained in this type of personnel selling are considerable, and have been widely used to gain control of entire markets. This approach usually warrants extensive public relations activities on the part of the company.

When mothers do not have regular contact with the infant-care medical establishment, personal selling to them directly in lieu of doctors may also increase the volume of sales. Usually, these company staff—'mothercraft nurses' or 'medical representatives'—will underlay their basic commercial role posing as bona-fide medical personnel by providing useful information and services to mothers. Their paramedical activities give the convincing appeal of medical or scientific advice to their recommendations on formula feeding. In the words of a marketing expert:

By getting over on the mothers' side of the fence the company becomes friend and counsellor rather than an advertiser trying to sell something.[60]

Detailpersons often try to reinforce the medical image by wearing uniforms, symbols of legitimate authority. Uniforms are often designed similarly to those of bona-fide medical personnel.

Personal visits, or demonstrations/presentations to groups of mothers are also commonly used. These are particularly effective among ill-educated groups that are not easily reached by other media.

In Africa, where many of the people cannot read, demonstrators do most of the initial selling. . . . These demonstrations are one of the key forms of communication in African countries. . . . We can't sell them on price offers. These people are the brand conscious people of the world, says Bassett, Borden's vice-president.[61]

The use of emotionally appealing materials such as coloured posters, slides, movies, and glossy pamphlets, as well as free samples and gifts are used to increase effectiveness of these selling actions.

Among illiterate populations, which still are to be found in many countries where Nestlé is in business, the most effective promotional message can be a poster depicting a pitcher of milk alongside a healthy-looking local child. Consumers dictate the (advertising) policy.[62]

Another activity with connotations similar to 'personal selling' is the organization of baby shows or contests, where free samples and the company's promotional materials are distributed.[6-64] Previous purchase of formula may be required to be allowed to participate.

All babies must be fully fed on Olac for at least three months immediately preceding the Baby Show.[65]

Although personal selling has many advantages over advertising promotion, it must be pointed out that detailing is quite expensive and that some physicians will refuse to see any detailperson. As a consequence, most firms have found a combination of both advertising and detailing to be the best solution.[55]

Types of communication used to promote sales of infant formula

At each stage of the development of an infant-formula market, the objectives of communication vary. The perceptions and attitudes of the target market towards the brand product change with time, developing along the path that goes from a mere awareness of the product to a final decision for regular purchase. Companies' communication strategies are therefore adaptive, tailoring their promotional activities progressively so that emphasis and types of communication correspond to the market development characteristic of the target market group. Table 12.15 presents a standard approach in the choice of relevant types of 'selling' communication.

TABLE 12.15   *A standard approach for choosing relevant types of communication*

Producer/seller's point of view, product life cycle	Phases in product adoption and main emphasis of communication process	Relevant types of communication
Introduction	Awareness	Announcements; classified ads.; slogans, jingles, teaser campaigns; descriptive copy
Growth	Liking	Billboards, competitive ads.; argumentation copy; image ads.;
Maturity	Preference	status–glamour appeals; demonstrations; shows
Saturation	Conviction	Point of sales display; retail ads.; special deals; give-aways
Decline	Purchase	Price offers; 'testimonials'; coupons-discount
	Cognitive dissonance removal	Argumentation copy; glamour/image ads.; testimonials; reminders

Adapted from reference 52

During the introduction of the product, the need is to create maximum awareness and knowledge about the new brand product, or the concept of its use. Experience has shown that during this phase, use of public announcements, ads., and popular jingles are particularly effective in attracting public attention. This kind of media advertising with mass outreach such as radio, public address systems, or even popular press, is precisely what is found in countries where traditions of breast feeding are still relatively strong and formula feeding being introduced, e.g. Kenya, Nigeria, Thailand, or Philippines. Appeals for the company's own product are selected and stressed, depending on their relevance to the basic motivations of the potential market, such as using the 'Western' appeal.

Then, to develop preference for the brand product, the types of relevant communication progressively shifts from description to argumentation. The argumentation copy will try to include the preference of the market by pointing out any fact, or subjective appeal, that may potentially lead the target consumer to the purchase decision. The study of consumers' attitudes towards the product, the brand name, and the company becomes central to the strategy. The objective is to eliminate possible resistances to the product and purchase act by appropriate messages. Then, overall attitude towards the brand image takes on importance. Communications focus on creating a decisive, positive disposition of the decider–buyer that will trigger adoption and purchase. The company's reputation and public goodwill are 'indispensable requirements if it is to thrive and achieve its objectives.'[55] Also, the social image of the brand product is important as it may transform formula product from a functional product, to which little cultural or social meaning is attached, to a status or prestige product. Then the product image is so developed as to affect the ego or self-image of the consumer, connoting membership to a 'higher' social group and increasing further aspiration to consume it.[54,55]

Many types of communication can be used to attain such objectives: image advertisements, billboards, demonstrations.[66-69] The objective of the messages is to reassure the prescriber/buyer of the excellent, superior properties of the product, and of the wisdom of the decision to purchase or prescribe the brand. The emphasis is the removal of cognitive dissonance. The repetition of positive image-building messages, 'reminder' advertisements help this removal of cognitive dissonance and crystallize the initial decision, keep brand loyalty high, and prevent 'drop-outs'.[55]

Infant milk is situated at different stages of its product market development in different countries. Accordingly, companies adopt practices that they see best fit the motivations and receptivity of the various target markets. These conditions vary between countries, and even within a country between different socioeconomic groups. So a whole range of communication techniques are used because each plays a specific role in the market development of the brand formula product. The importance of such activities in promoting formula feeding and bottle use cannot be understood by examining each of them separately. They need to be studied, not in isolation, but within the overall logic of the companies' communication strategies.

Role of advertising. In the framework presented above the role played by certain promotion practices, such as advertising, in the overall marketing strategies of the companies can be better understood. Advertising to the public, through mass media, is likely to be particularly important in the early phases of the brand product life in a given market. The crucial role of paediatricians and maternity wards in product adoption and purchase decision also suggests that such public advertising activities are more likely to happen in areas where medical coverage is lacking.

For this reason, most mass media campaigns for infant formula appear to be in developing countries, and more specifically directed towards the low-income urban groups and, to a lesser extent, rural groups, that sill have little regular access to health services. Various studies of advertising in the developing countries through the press or radio indicate a tendency for the companies to advertise selectively for these low-income groups.[32,67,68]

In these situations, the use of popular jingles and slogans corresponds to the phase of familiarization to the product name, crucial in its market development. For the most traditional environments it may even constitute an introduction to the concept of formula feeding itself. In this case, it is important to note that there will be, by definition, a period where large advertising media campaigns will be observed jointly with widespread practices of prolonged breast feeding. Thus, testimony given by health professionals that there is currently breast feeding in a particular country does not preclude its imminent decline in the absence of specific protective/preventive regulations and policies.

The lack of apparent effect is misleading, because it is the purpose of these campaigns to pave the way for a gradual acceptance of the product and prepare the ground to move on to the next step in the product life cycle. The coincidence of continued breast-feeding practices with intensive mass media advertising, such as has been observed in Africa or Asia, should therefore leave no doubt about the persuasive effects they have on future infant-feeding practices. These are indeed the precursors to formula market development.

For instance, a survey in Lagos (Nigeria) indicated that about two-thirds of the mothers artificially feeding their babies did so not because of work, or lactation failure, but because they got the idea from the milk advertisements on the radio, and believed it was good for the baby because they had seen some well-to-do mothers feed their babies artifically.

Mass-media advertising has been considerable in some countries. In 1973 in Kenya a three-week monitoring of radio advertising revealed that Nestlé's Lactogen accounted for 11 per cent of all Swahili radio advertising time.[32] In the Philippines, all three major companies used newspapers, magazines, radio, and TV in 1975–76 for their 'selling' communication efforts.[31]

Advertising to the medical profession fulfils a role somewhat different than that of public advertising. For the medical circles, the market for almost all formula products is already saturated, in the declining phase of the product life cycle. Communications then strive essentially for removal of cognitive dissonance, and to act as a reminder for 'repeat' prescription pattern.

A central element that mediates physicians' prescription habits is their perception of the risk benefits in prescribing a particular product (and the satisfaction of his client), as well as their subjective confidence in estimating the effects of its usage: 'An internal debate is constantly being waged by the physician. He must justify his decision to his own value system.'[55] At the present 'maturity' stage obtained by most infant-formula products, advertising in professional journals or promotional literature hardly conveys new substantive information. Rather the role of advertising at this stage is precisely to confirm paediatricians in their routine recommendations for artificial infant-feeding patterns and choice of specific formula.

Professional journals function as a legitimizing channel and influence prescribing habits by bestowing some form of approval to the physicians' use of the product.[70]

The journal ads take on a certain psychological aura of authority by running cheek by jowl with scientific and expert editorial matters.[71]

Rather than informing technically sophisticated audiences about dramatically changing products, these advertisements are additional ways of engraining the physicians' repeat prescribing patterns. A positive brand image is useful in obtaining this behaviour and justifies high levels of advertising.[72] The same could be said for the numerous favours given by companies to the nurses, nutritionists, or paediatricians and their associations.[57,58]

Use of motivation in 'selling' communications

Since motivations determine the ultimate decision to prescribe or to buy a particular product or brand, their identification is imperative in the success of the selling communication process.

With respect to the promotion of infant formula, we have mentioned that different types of communications are better adapted to each stage of the product life cycle. The messages that are most likely to influence the 'formula' decision have also been identified by industry. An examination of the advertising copy of formula promotion materials illustrates the recurrence of those themes and factors most effective for selling messages to affect consumers' behaviour. They prey generally upon the basic motivations related to infant feed: desire for infant strength, good health, and nutrition. But in addition they also attempt to remove possible elements of resistance that may be associated with the product and anxiety relative to its legitimacy, safety, or ease of use.

Particular mention should be made of the appeal to the 'almost equivalence' of the product with breast milk. It is one of the most commonly used messages in promotion, and is always part of the label/labelling communication. Presented under various formulations, reference to breast-milk constitutes certainly one of the most powerful selling arguments, although evidence has shown that breast feeding and artificial feeding are usually not equivalent processes for the infant health.[46]

A few other themes consistently used in communication comprise:[67-72]

1 Joy, love, and affection for the baby.
2 Good health and nutrition, strength, and happiness of the baby resulting from product use.

3 Superior quality of the product, including reference to breast milk and scientific arguments, but also taste and biological purity.
4 Dependability, safety, and facility of product use.
5 Physician- or medical-related appeals—'testimonies'.
6 Manufacturer-related appeals—biological purity.
7 Social-status-related appeals—'high class', 'modern'.

The creative execution corresponding to these basic messages depends on the nature of the target market. In countries where high rates of illiteracy prevail, advertising (to be effective) must stress symbolic representations, especially with appeal to popularized jingles, brand names, colours, and logotypes.[72]

The reliance on imagery of beaming, healthy babies is almost universal in the labelling of the cans. Copy emphasizes 'a subtle approach, in which the tone and atmosphere is one of human tenderness, while illustrating the health and strength achieved as a result of the brand's use'.[72] There is little doubt that such efforts are intended to succeed in orienting mothers' infant-feeding choice towards the use of a given product, and that indeed they manage to do so.

The important role of brand identification needs also to be mentioned. Appeals will be all the more effective if they promote a product image that figures 'high' in the consumers' minds. Although formula could be classified primarily as a functional product, the product can also affect the ego or self-image of the mother. To that extent, formula feeding may become a prestige or status behaviour, either denoting leadership or membership to a certain local group. This is generally true in developing countries where formula use and bottle feeding is associated with modernized élites, and sets the pace of emulative behaviour. Formula becomes a product which serves the function of identifying or extending the mothers' ego in a direction consistent with social aspirations.[66] Studies in India illustrated that there was an inverse relation between the quality image of branded baby foods and family income.[84]

Although distribution of free samples to hospitals, doctors, and mothers has been considered an accepted practice by the industry, it has a most profound adverse influence on breast feeding. Besides contributing an obvious encouragement to formula or bottle feed the infant since the moment of birth, it is also a major element contributing to price incentive in the early and decisive introduction of the breast-milk substitute.

Regular distribution of free samples contributes to the lack of price awareness on the part of health professionals. They may not become aware of the costs of products they prescribe when they receive free formula supplies from the manufacturers. Company gifts to health personnel may induce them to establish a projection from their own babies to their clients. This may be most misleading, according to the late President of the Rockefeller Foundation, Dr J. Knowles. In general, doctors belong to the highest income group and do not share the living conditions of the majority of the population.[73]

The provision of formula products at reduced price, or free of charge through national health insurance schemes, also tends to induce a necessity effect augmenting disregard for the real costs of artificial feeding by some consumers.[24] Such subsidies may facilitate the access to formula by the consumers, as well as encourage formula prescription and recommendation habits by paediatricians. But in many developing countries, where such programme exist, the overall effects are negative since usually only a small percentage of the population is eligible to participate in such programmes, and professionals are none the less influenced to recommend a standard breast-milk substitute feeding for all patients, regardless of their socio-economic background. Many maternity clinics adopt uniform infant-feeding routines; therefore, these programmes help in establishing a model for a feeding behaviour that may be inadequate for the majority of the infant populations, be it urban or rural. (An interesting illustration is found in the letters from doctors in the Philippines, which on the one hand were acknowledging the widespread use of companies formula samples in their wards, and on the other hand were claiming that mothers could not afford the formula products.)[74]

In fact any free or discounted supply made available by the companies is highly delusive, since compensatory revenues from price and sales increases can be obtained from the commercial market. Somehow, consumers have to pay for it:

Sampling promotes "prescription" sales. Pharmaceutical sampling is seed sown for the sole purpose of raising up a bumper crop of prescription sales . . . on a successful product, increased tempo of sampling operations will push sales up further. . . . Only 10 per cent of doctors pass on samples to patients without being prescribers.[71]

Conclusions

Codes of conduct and Government intervention

For a number of years, commercial health professionals have called for restraints on the marketing activities of the infant-formula industry because of their pervasive effect on infant-feeding practices. Since the early 1970s, United Nations experts' meetings where industry representatives took an active part have focused on the problems of infant nutrition and invariably issued strong and unambiguous resolutions that formula promotion activities discouraging breast feeding should be discontinued in developing countries.[75]

Industry took some steps to amend its marketing behaviour; but the extent of its genuine concern is indicated by the fact that half a decade later, no one company could report the implementation of any system for determining potential negative effects of use of their products of adoption of bottle feeding.[31] At the same time, companies acknowledge that they maintain elaborate and costly systems of market research as a basis for their product development and promotion plans.

It seems that effective pressure on the companies to alter their promotional programmes came not through the medical channels, despite these efforts, but when concern of the public at large was also brought to bear. Starting in the mid-1970s, a number of promotional activities were exposed to public opinion mounting criticism was voiced in Europe and the USA about the legitimacy of such activities.[67] Largely in response to public concern, codes of conduct for marketing were finally formulated by the companies.[75]

If the companies have come to recognize publicly that their previous promotion activities needed to be revised, their response fell generally short of a significant reconsideration of the way in which they conduct business and promote formula products. The restraints to promotion that have been adopted in the different 'codes of ethics' are actually of such modest consequence that they appear to be more aimed at defusing public criticism than at a serious self-questioning of the legitimacy of their promotional activities.

In particular, the 'code of ethics' adopted by the members of the International Council of the Infant Food Industry has been criticized for this reason.[76] The Council had set restraints, for instance, on what types of uniforms the mothercraft personnel should wear, when it is the existence of such sales personnel itself that is questionable. In the same spirit, the companies' revived recognition of the superiority of the mothers' milk has been transformed into yet another commercial appeal for the breast-milk substitute products.[40]

There are differences, however, among the codes that have been established by ICIFI and individual companies[75] who, by their own account, acknowledge that changes are possible only because they correspond to minor modifications of their own mode of operations. They are acceptable because they were already part of or consistent with their respective marketing strategy, and do not alter their basic competitive position in the market. For instance, pharmaceutical companies—which have traditionally concentrated their promotional efforts towards the medical profession through 'ethical' marketing—are more willing to give up the types of communication oriented to the public, whereas the food companies refuse to accept such a limitation.

When significant changes are proposed that would reduce the companies' influence on infant-feeding practices, such as stopping the distribution of free samples or the establishment of a special in-house committee to review breast-milk substitutes marketing policies and practices, these measures are strongly opposed because they would affect the companies' ability to create new consumers and gain control of target markets:

Changes called by the shareholders' proxy would result in the discontinuation of free samples, use of mothercraft personnel, and promotion of infant formula within the medical profession. At the time when the company has completed in part or contemplates new or expanded facilities for infant formula in Ireland, United Kingdom, Japan, South Africa, Australia, The Philippines and Colombia, these changes in the company's marketing policies would *greatly inhibit* the sales of the products of these new or expanded plants.[77]

Recent experience suggests that codes of conduct elaborated by the industry and adopted on a voluntary basis may constitute a first step towards the reform of the promotion of breast-milk substitutes. But they have brought modifications which remain secondary to main promotional thrust. In addition, it appears that compliance to the restraints subscribed to by the companies in the codes is dependent on the existence of an independent monitoring system that could exert continuous surveillance of the individual company's marketing behaviour. The companies have varied behaviour from one country to another depending on the amount of public attention that has been brought on their promotional activities. Promotion to the public has been reduced in Nigeria and Kenya after disclosure of massive promotional efforts and analysis of the effects there, whereas they continue unabated in other countries such as Malaysia, Philippines, or the Yemen Arab Republic.[31,78-80] Unless effective pressure can be kept on the companies to make them conform to their own guidelines, such as through close monitoring, disclosure of marketing activities, and communication of violations to the public, market forces will be stronger than the agreed-upon rules.

The breast-milk substitutes and the cigarette industries appear similar. Both depend on large promotional efforts while consumption of the product is hazardous to health. In the 1960s, public attention in the USA was aimed at health hazards. It prompted the nine major tobacco companies to band together in a council and to adopt a voluntary code of ethics for advertising. With the relative diminution of public scrutiny, and under competitive pressures not unlike those existing in LDCs for infant formula, the code degenerated, and 5 years later only two companies declared that they were still observing the code stipulations. Almost none or little of the restrictions that have been mandatorily imposed on the transnational tobacco corporations in their home countries have been followed by their affiliates in developing countries, nor were they observing the WHO recommendatory restraints.[81]

The cigarette industry code was in many respects similar to the ICIFI code. It dealt not with the ban of certain questionable promotion practices but rather how the practices should be adapted. Continued justification was found for the most adverse forms of promotion, such as public advertising, by declaring the intention to focus it only to certain target groups among the population. A similar approach had been also recommended in the case of formulas.

A series of reports of the Federal Trade Commission (USA) on the subject year after year documented the ineffectiveness of the cigarette code. In the following quotes the similarity to the formula promotion issue is emphasized by the parenthetical additions:

On their face, the various advertising codes may appear to set proper guidelines for cigarettes (formula) advertising (promotion). But in practice, it is impossible for cigarettes (formula) manufacturers to comply with the codes without making known the health hazards of smoking (bottle feeding) or diminishing in any way the appeal of their promotional efforts.[82]

Similar also is the market 'segmentation' approach:

Promotional materials that have an impact on the adults (rich people) cannot be assumed to leave unaffected the less than 21 years old (low-income group). The world depicted in promotional materials very often is a world to which the young (poor people) aspire. To them, smoking (bottle feeding) may seem to be an important step towards and a help in growth from adolescence (backwardness) to maturity (modernity).[82]

Such conclusion could well apply to the breast-milk substitute markets. Although it is clear that formula is needed for certain groups of babies unable to breast feed early in life, it is not possible to argue for double standards and to promote adoption of bottle feeding for the privileged, richer part of the population, while recommending breast feeding for the 'poor'.

It is certainly no safeguard to promote bottle feeding exclusively upmarket, for then poor people see it as evidence of proper and better thing to do.[72]

Lessons can be drawn from the cigarette experience. There is an indication that no voluntary code will be effective without legislative follow-up actions. Numerous instances of violations of the codes on breast-milk substitutes marketing have been gathered around the world since their adoption.[78-80] It is likely that an effective and lasting curb on adverse breast-milk substitutes promotion will be possible only through appropriate formalization of the changes necessary to safeguard breast-feeding practices, both nationally and internationally.

At the present moment, when a consensus seems to emerge from industry to medical and public health experts, it is important that such a consensus be institutionalized into a durable (reference) document. An 'International Code of Conduct', upon general acceptance by the interested parties, may serve to set the principles for international regulations to be further complemented by specific measures in each country's legislation.

The need for this action is particularly important given the increasing pressures of increased competition and escalating promotional efforts on the health systems and the public of developing countries.

International code of conduct and regulations by Governments

Analysis of the business outlook for the formula industry points in the direction of heightened competition among the companies. The powerful forces at work in this process directly undermine existing breast-feeding practices. Because market shares are directly derived from certain types of promotional activities and their intensity, with an increase in competition, these activities are stepped up at the expense of support for the more appropriate breast feeding of infants. Activities most destructive to breast-feeding practice are also those most crucial in gaining control of breast-milk substitutes markets. Companies are understandably unwilling to grant concessions in this area. Restrictions in these practices would damage the ability of a company to achieve its corporate growth objectives. Soon the voluntary marketing restraints or codes agreed on by the companies under outside pressures will be challenged by management. As a senior executive officer of a large US company puts it: 'Our promotion practices are in fact determined by the intense competiton which exists among manufacturers of infant formula products.'

In such circumstances, as long as the restrictions on promotion are not put in to a regulatory context and applied to the whole industry, there is little chance that the companies will conform in a lasting way to measures that would limit their direct influence on infant-feeding practices. What they come to admit and agree to under the combined, intense pressure of scientific and public health professionals and the public will deteriorate rapidly when the momentum of these efforts subsides.

In summary the emergence of industry voluntary codes of ethics should be interpreted as the willinginess from the industry to acknowledge that it had been led to undesirable and noxious promotion practices in the absence of Government regulations. As a consensus emerges on legitimate restrictions to certain practices, in view of their undermining of breast feeding, these restrictions should be rapidly formalized, e.g. bans on mothercraft personnel, public advertising, free samples, and other donations; elimination of commissions on sales; restriction of access to health care facilities for promotional purposes; communications to health professionals limited to scientific factual matters.[83]

Government regulations are needed to avoid any degeneration of the voluntary codes and attempts to circumvent them, because when this happens, even for a single company, there is necessarily an escalation from the other companies and the policing effect vanishes quickly. So far, companies appear to have had difficulties complying with the codes they have themselves devised. The existence of binding regulations facilitate the companies' observance of principles they may value. On principle, and because it obviously limits the range of their promotional activities, the companies are vehemently opposed to any regulation and Government interventions. But in fact, even when they have argued that such measures were not justified from their own point of view, once passed into laws the rules of the game have changed for all of them. The companies then accept the new rules as a part of their strategy.

The elaboration of an International Code of Conduct for the promotion of formulas is an important step forward. In developed countries, adoption of measures codifying the distribution of free samples has run into considerable opposition from the industry.[85,86] It is conceivable that lobbying in developing countries will be even more powerful to stall any legislation regulating the industry practices or made then innocuous. In this respect, the existence of an internationally recognized code is needed to support public health officials and Government efforts in many countries.

As outlined by the UN Centre on Transnational Corporations, several strategies are available to Governments to protect consumers from abusive pharmaceutical promotion:[87]

1 By slowing commercial promotion together with State Counterpromotion –competition.
2 By permitting commercial promotion, but regulating it.
3 By prohibiting commercial promotion, and/or make it a function of the State.

In the case of infant formula, the first strategy has proved to be fully inadequate. Health services cannot compete with the amount and sophistication of efforts devoted by the companies to clear the way for adoption of bottle-formula feeding by the population. Sound programmes for maternal and infant nutrition—including appropriate support for breast-feeding practices—cannot be undertaken successfully by the health services without eliminating the larger part of present promotional efforts of the formula producers.

The second strategy would attempt to regulate the promotional activities, but loopholes are always open, especially if the regulations are limited to indicating the way certain activities should be undertaken.[78,84] A ban of those activities already mentioned above provides an easier basis for implementation and enforcement than measures requiring a qualitative assessment of compliance.

The third alternative gives the State responsibility for promotion/education and distribution of the breast milk substitute products. This has been adopted by several countries, even for other foods.[88] Competitive bidding, with diversified suppliers and efforts to develop a national/regional production capacity of generic formula are other aspects of the issue that have been considered.[88]

Depending on local circumstances, individual countries may opt for more or less regulations and participation of the State administration. Until developing countries have developed enforcements procedures and an administrative capability, it seems that an international commission is needed to provide adequate support for the policies undertaken by the national Governments. This commission would also have a role to play in the compliance of the companies with the regulations.

Without Government intervention through regulations or direct control of formula promotion, the evidence suggests that the present trend towards the decline of breast feeding will accelerate—at great economic, social, and human costs. The promotional efforts of the formula companies are so pervasive as to make it very hard for health workers to develop effectively the preventive health and nutrition programmes appropriate for the resources and living conditions of the majority of the world population. In particular, it seems impossible to engage in programmes supporting breast feeding with any chance of lasting success in the absence of such Government policies and international regulation.

Postscript

At the October 1979 Joint WHO–UNICEF Meeting on Infant and Young Child Feeding, a consensus was reached between industry representatives and all other concerned parties.[89] The recommendations of the meeting were later unanimously approved by the 33rd World Health Assembly in May 1980, and it was further decided to prepare in collaboration with industry, health, professionals, Governments, and consumer representatives an International code.[90] In May 1981, the 34th World Health Assembly adopted, with a lone dissenter vote by the USA, the recommendatory version of an International Code of Marketing Breastmilk Substitutes.[91] While this document provides a useful basis for further regulatory action to be taken in each country, its recommendatory nature makes it inherently a weak instrument.

A number of activities by the leading firms, intended to undermine the development and adoption of the code,[5-7,92] bode ill for its future effectiveness, unless mandatory controls and adequate monitoring measures are adopted in host countries and internationally. It is now apparent that concerted industry pressures were behind the USA's sole *no* vote against the WHO code. The leading breast-milk substitute firms have demonstrated a powerful ability to override the recommendations of concerned technical agencies in the USA. To foster their corporate objectives, US infant-formula manufacturers engaged in a 'systematic campaign of disinformation intended to create a USA Government position against the proposed code', according to Dr Stephen Joseph, the highest-ranking health professional at the United States Agency for International Development.[92] Recent disclosures have shown the ability of the infant-formula lobby to have what Dr Joseph called its 'self-interested arguments'[92] prevail over scientific and public health considerations, in a country where these considerations have been thoroughly documented.[34,49] Such corporate behaviour is a clear indication of the extent to which these firms may be willing to go to pursue their objectives.

At the same time, while the effectiveness of the international code hinges on the adoption of follow-up legislation in each country, the industry leaders have also stepped up pressure upon host Governments for the adoption of weak and voluntary local codes.[93]

Leading breast-milk substitute corporations have used their power to elicit specific policy decisions from Governments to promote their interests, in spite of more than one million infant deaths annually that may be at stake in the 1980s, according to UNICEF's Executive Director. It is clear that unless public health professions in each country take on an active responsibility to ensure that effective regulations and control of corporate behaviour are adopted and implemented, little can be expected from the new International Code of Marketing, however well intentioned it may be.

Acknowledgements

I thank A. Domike, M. C. Latham, and D. G. Sisler for stimulating comments on some of the ideas developed in this chapter. The views that are expressed do not necessarily represent those of the United Nations.

References

1. Jelliffe, D. B. and Jelliffe, E. F. P. *Pediatrics* **66**, 637 (1980).
2. *Advertising Age* Feb. 2 (1981).
3. Nutritional Committee of Canadian Paediatric Society, and the Committee on Nutrition (US). *Pediatrics* **62**, 591 (1978).
4. WHO/UNICEF. *Infant and young child feeding. Current issues.* WHO, Geneva (1978).
5. Saunders, Memorandum to Mr Fürhrer,CEO, Nestlé. 1980, August. Vevey; Mintz M., *Washington Post*, Jan. 4 (1981).
6. Thirty-Fourth World Health Assembly. *Item 23.2 Provisional Agenda*, WHO, Geneva, A34/INF DOC./10 (1981).
7. Senate Foreign Relations Committee—Hearings on Mr E. Lefever's Nomination as Assistant Secretary of State for Human Rights. United States Congress, 1981, June, Washington DC *New York Times* 31 May (1981).
8. Abbott Laboratories. *Annual report* (1979).
9. *Financial Times*. Dec. 15. London (1980).
10. *International Daily News* Feb. 4. London (1981).
11. James, B. G. *The future of the multinational pharmaceutical industry by 1990.* Wiley, New York (1977).
12. *Scrip*, No. 506. July 6, p. 9 (1980).
13. Scherrer, F. M. *Industrial market structure and economic performance.* Chicago, Rand McNally (1980).
14. Steele, H. *New Physician* March (1971).
15. Walker, H. *Market power and price levels in the ethical drug industry.* Indiana University Press, Bloomington (1971).
16. Markham, J. *Economic incentives in the drug industry.* Johns Hopkins University Press, Baltimore, Md. (1969).
17. Brooke, P. A. *Resistant prices.* Ballinger, New York (1975).
18. McEvilla, J. D. In *Principles of Pharmaceutical marketing* (eds. Keller and Smith). Baltimore, Md. (1969).
19. Mueller, W. F. and Rogers, R. T. *Review of economics and statistics*, 89–96 (1981).
20. Ward, R. W. and Behr, R. M. *Am. J. Agric. Econ.* 113–17 (1980).
21. Slatter, S. P. *Competition and marketing strategies in the pharmaceutical industry.* Croom Helm, London (1978).
22. Reekie, W. D. *Barriers to entry and competition. Economics of innovation in the pharmaceutical industry.* Association of British Pharmaceutical Industry, London (1969).
23. Mueller, W. F. *Competitive problems in the drug industry.* Subcommittee on Monopoly. United States Congress. US Government Printing Office, Washington DC, pp. 1806–61 (1968).
24. Reekie, W. D. *The economic of the pharmaceutical industry.* MacMillan, London (1975).
25. Federal Trade Commission. *Economic report on the influence of market structure on the profit performance of food manufacturing companies,* US Government Printing Office, Washington DC (1969).
26. Weiss, L. The concentration-profits relationships and anti-trust. In *Industrial concentration: the new learning* (ed. H. Goldschmidt). Little Brown, Boston (1974).

27. Connor, J. M. and Mueller, W. *Market power and profitability of US multi-national corporations in Mexico and Brazil*. US Senate Subcommittee on Foreign Relations. US Government Printing Office, Washington, DC (1977).
28. Wells, L. T. A product life cycle for international trade. In *International marketing strategy* (ed. H. Thorelli and H. Becker). Pergamon Press, New York (1980).
29. Post, J. E. and Baer, E. C. *Res. corporate soc. performance and policy*. **2**, 157 (1980).
30. M and R Laboratories; Mead Johnson. *Annual reports* (1955-1965).
31. Wickstrom, G. *Infant food marketing study*. WHO/CIE Collaborative Study WHO/CIE, Geneva, Goteborg (1978).
32. Langdon, S. *Rev. Afr. Pol.* **2** (1975).
33. Kaplinski, R. Review. *Afr. Pol. Econ.* **14**, 90 (1980).
34. Post, J. E. In *The marketing and promotion of infant formula in the developing nations*. Senate Subcommittee on Health and Scientific Research. US Congress, Washington DC, pp. 120-216 (1978).
35. Fomon, S. J. *Infant nutrition*. 2nd edn Saunders, Philadelphia (1974).
36. *Studies in Family Planning*. *Stud. Fam. Planning, Special issue*, **12**, 4 (1981).
37. Baker Laboratories. *Civil action 75-671 against Ross Laboratories and Mead Johnson*. US District Court, Pennsylvania (1975).
38. *L'expansion*. June. Paris (1972).
39. Ross Laboratories *Detail persons training manual: hospital sales*. Columbus, Ohio (1975).
40. Borgoltz, P. *Economic and business aspects of infant formula promotion: implications for nutrition policies in developing countries*. United Nations Geneva (1979).
41. Syntex Laboratories. *10-K Report* (1979).
42. *Financial Times*, Dec. 15 (1980).
43. *Business Week*, Feb. 2, p. 56 (1981).
44. *Advertising Age*, Feb. 2 (1981).
45. *Indian Express*, Jan. 2 (1981).
46. Jelliffe, D. E. and Jelliffe, E. F. P. *Human milk in the modern world*. Oxford University Press (1978).
47. Carlson, S. *Dev. Psychol.* **11**, 143 (1978).
48. Hales, D. J. *Dev. Med. Child Neurol.* **19**, 454 (1977).
49. Bergevin, Y. *et al. Ambulatory pediatric association program and abstracts*. San Francisco. April 30-May 1 (1981).
50. Winikoff, B. and Baer, E. *Am. J. Obstet. Gynecol.* **138**, 105-17 (1980).
51. Isenalumbe, A. *Study of the relationship between breastfeeding and reliance on four specific sources of information on infant feeding practices*. Ph.D. Thesis, New York University (1979).
52. Majaro, S. *International marketing*. Alan & Unwin, London (1977).
53. Harrell, G. D. Pharmaceutical marketing. The *pharmaceutical industry* (ed. C. Lindsay). Wiley, New York (1978).
54. Windham, D. *The pharmaceutical industry*. London (1967).
55. Smith, M. C. *Principles of pharmaceutical marketing*. Lea and Febiger, Baltimore, Md. (1968).
56. Ross Laboratories. *Annual report* (1962).
57. Reports Center for Science in the Public Interest. Washington, May, p. 5 (1981).
58. *Calif. Pediatr.* Winter, p. 9 (1981).
59. Moyers, W. *CBS Reports. Transcript*. New York, July 5 (1978).
60. *Printers Ink*. Nov. 25 (1960).
61. *Printers Ink*. June 13 (1968).
62. *Business Abroad*. Where sales begin. Nestlé's nutrition clinics. June (1970).

63. *Ufusan Konsumer*. Dec. (1978).
64. *Business Abroad*. June (1970).
65. *Daily Gleaner*. Advertisement for the 11th Anniversary, Mead Johnson Olac Baky Show. Jamaica, June 16 (1971).
66. *Business International*. Aug. 6 (1978).
67. Chetley A. *The baby killer scandal*. War on Want, London (1979).
68. Greiner, T. H. *The promotion of bottlefeeding by multinational corporations: how advertising and the health professions have contributed*. Monograph Series No. 2. Cornell University, Ithica, NY (1975).
69. Greiner, T. H. *Regulations and education: strategies for solving the bottlefeeding problem*. Monograph Series No. 4. Cornell University, Ithaca, NY (1977).
70. Worthen, D. B. *Br. J. med. Educ.* **7**, 109 (1973).
71. NPC. *Proceedings: pharmacy educational forum*. Princeton University Press (1959).
72. Sethi, P. *Advanced cases in multinational business operation*. Good Year Publishing (1972).
73. Knowles, P. *PAG Bull.* **7**, 3–4 (1976).
74. Bristol Myers. *Additional materials in marketing and promotion of infant formula*. Committee on Health and Scientific Research US Congress. US Government Printing Office, Washington, DC, pp. 1087–1928 (1978).
75. Baer, E. C. and Post, J. E. *Rev. Int. Commission Jurists*. **25**, 52–61 (1980).
76. International Council of Infant Food Industries (ICIFI). *Code of ethics and professional standards for advertising*. Product Information and Advisory Services for Breastmilk Substitutes, Zurich (1975).
77. American Home Products. *Statement to shareholders*. 27 Jan. (1977).
78. Consumer Association of Malaysia. *The other baby killer*. Penang Mar. (1981).
79. *Infant Formula Promotion 1980*. IFBAN, London (1981).
80. Firebrace, J. *Infant feeding in the Yemen Arab Republic*. Catholic Institute for International Relations/War on Want, London (1981).
81. Wickström B. *Cigarette marketing in the Third World*. *Göteborg University (1979)*.
82. *Federal Trade Commission. Report of the commissioner*. US Congress. US Government Printing Office, Washington, DC (1967).
83. WHO/UNICEF. *Background document. Joint meeting on infant and young child feeding*, section 2–5. WHO/UNICEF, Geneva (1979).
84. Medawar, C. *Insult or injury?* Social Audit, London, p. 92 (1978).
85. D and C Industries. November (1976).
86. *Med. Mkt.* Dec. (1976).
87. Center on Transnational Corporations. *Transnational corporations in the pharmaceutical industry*. United Nations, New York, ST–CTC–9 (1979).
88. Baer, E. C. *Stud. Family Plann. Special Issue.* **12**, 4 (1980).
89. Joint WHO/UNICEF Meeting on Infant and Young Child Feeding. *Statement, recommendations. List of participants*. WHO/UNICEF, Geneva, (1979).
90. Thirty-third World Health Assembly. *Res. WHA* 33.32. 1980, Geneva (1980).
91. Thirty-Fourth World Health Assembly. *International code of marketing of breastmilk substitutes*. WHO, Geneva. Resolution 34.33. 21 May (1981).
92. Joseph, S. *Statement in opposition to the US Government position on the WHO/UNICEF international code on the marketing of breastmilk substitutes*. American Public Health Association, Washington (1981).
93. *Times of India*, 19 January 1981 (1981).
94. Grant, J. *Executive Director UNICEF, UN Conference on Least Developed Countries, Paris* (1981).
95. Martinez, J. and Maliezienski, J., *Pediatr.* **60**, 260 (1981).

13 The precipitous decline in breast feeding in the Gulf Coast States

JAMAL K. HARFOUCHE

Introduction

With the exception of the two Yemens and Oman, the Gulf Coast States considered in this paper have the highest per capita income and are the wealthiest countries in the world. They are characterized by rapid socioeconomic change uprooting traditional bedouin lifestyles, extensive population movement with heavy influx of foreigners (particularly males in the labour force), urbanization, modernism, transfer of complex and costly technology, prestigious hospitals, and minimum concern with essential health care.

The precipitous decline in breast feeding as a unique phenomenon in the Gulf States can be appreciated only if viewed in historical perspective. The drastic change in the incidence of breast feeding which took about a century to reach its lowest ebb in the industrialized countries is now taking place in the Gulf States in less than 10–15 years. In no time 'money', 'modernity', and 'commercial advertising'—an interlocking complex of factors promoting bottle feeding—have usurped the Koranic tradition of breast feeding for 2 years (Al-Baqarah: 233) and the infant-feeding principles contained in the 'Canon of Medicine' by Avicenna, faithfully observed since the eleventh century; and have radically modified the cultural image of the female, whose vital biological role—'lactation', the symbol of 'motherhood' and 'mothering'—now appears almost to be a legend of the past.

Supportive evidence

Infant feeding practices in the Gulf States have been scarcely studied[1] and little information has, thus far, appeared in the literature. Recently, data collected by special surveys have appeared in unpublished reports and theses, providing adequate evidence concerning the rapid infiltration of artificial baby foods and feeding bottles, associated with a precipitous decline in the incidence and duration of breast feeding.

Bahrain

In a study of food habits in urban and rural areas, Musaiger[2] found in 1977 that many mothers had abandoned traditional feeding practices in favour of bottle feeding and early weaning, and that the higher the socioeconomic status of mothers, the shorter the duration of breast feeding.

In 1977, at least ten different varieties of imported baby foods were on the market,[3] and the Ministry of Health distributed some of them (i.e. Lactogén, Namex, Nestogén, half- and full-cream, Pelargan, Similac, S–26, and Farlene) through the different MCH and health centres in the country, costing in 1978 a total of B. Dinar 29 592 719. Evidence pointed to a close link between this measure and the declining incidence of breast feeding and the prevalence of infantile marasmus, secondary to gastroenteritis associated with bottle feeding.

In the period 10 August–13 October 1977, 746 culture-confirmed cases of *Vibrio cholera*, biotype El Tor, serotype Ogawa, occurred in Bahrain.[4] Cases appeared throughout the country and affected mostly infants (less than 1 year of age), young children, and working men; eighty-one culture-confirmed infants with symptoms were identified (attack rate 84 per 10 000). The highest age-specific attack rate (125 per 10 000) was seen in the 6–11-month age group; 74 per cent of the infants were admitted to hospital, and one 6-week-old infant died.

A matched-pair case–control study was done to examine the effect of different feeding practices, weaning, and maternal cholera infection on clinical cholera infection in fifty-five of the sixty-one culture-confirmed infants (all aged less than 1 year), and a successful interview was completed for forty-two matched pairs (age groups 0–2, 3–6, and 7–11 months). The study revealed a highly significant association ($P = 0.0004$) between cholera infection with symptoms and history of being principally bottle fed. Infants exclusively bottle fed had a relative risk of 7.0 for cholera infection with symptoms, compared with those exclusively breast fed. Significantly more cases than controls ($P<0.02$) had been weaned from breast feeding before becoming ill.

In Saudi Arabia, where infant feeding practices are similar to those in Bahrain, a report of a recent outbreak also showed a high incidence of cholera for infants under 1 year old.[4]

Democratic Republic of Yemen (PDRY)

Clinical Family Planning Services in PDRY started at the MCH/FP Centre of Al-Gamhouria Hospital, Aden, in August 1976.[5] The Directorate of MCH/FP decided to study the characteristics of the first 1000 mothers to provide a baseline for purposes of future planning and evaluation. Being the first investigation of its kind, a long list of variables was studied, including breast-feeding practices among mothers who had children below the age of 18 months (670 out of 1000 mothers; 87 per cent of the children were below the age of 12 months and 60 per cent were less than 4 weeks through to 5 months). The majority of mothers were 20–29 years of age; 75.9 per cent were housewives and the remainder had some kind of gainful employment.

Individual clinical cards of the mothers and records of the FP clinic were used as sources of data. The results revealed that only 11 per cent of the mothers were exclusively breast feeding their babies at the study period;

34 per cent breast fed partially; and 55 per cent were bottle feeding. Since very few mothers were breast feeding their babies, the Directorate of MCH/FP recommended that all the MCH centres be urged to give regular talks on breast feeding in the centres as well as during home-visits, and that all mass media be utilized to inform the public about its advantages.

Moreover, in February 1978 a nutritional survey was undertaken in two villages in the Third Governorate as a pilot study;[6] 399 infants 0–3 years of age were included in the study (250 from Laboos in the mountainous area of the western province at about 2100 m above sea level; and 149 from Mahfed on a plateau in the eastern province, 60 m above sea level, forming approximately 4 and 8 per cent, repsectively, of the corresponding child population). The median age of the mothers was 24.9 years; only 4 per cent of them had passed 1 year or more in school; 18 per cent of all investigated children were the only living child of the family; and 85 per cent of them were born at home.

In both villages cows' milk products for infants were seen in abundance. *Dutch Baby* was sold everywhere. Dilution of preparations was prescribed in English and detailed with instructions neither readable nor realizable by the village women. Bottles with formulae and/or gruels based on cows' milk were introduced very early, with a consistent trend of earlier and more extensive use of bottles in Mahfed than in Laboos. Before the age of 6 months more than 50 per cent of the infants were bottle fed, apparently in the form of 'mixed feeding' (i.e. fed both breast milk and formula/gruel). A 55 per cent prevalence rate of protein-energy malnutrition (PEM) was found, increasing in both incidence and severity with age, and had not reached a maximum by the age limit of the study (i.e. 3 years).

Kuwait

In March 1978, the newly established Nutrition Unit[7] conducted a pilot study on a sample of 509 (245 Kuwaitis and 264 non-Kuwaitis) to investigate the feeding practices of infants (0–2 years) attending the Preventive Medicine Centre—Farwania, Fayha, and Madina. The results indicated that 86.1 per cent of the Kuwaitis and 75 per cent of the non-Kuwaitis were artificially fed. The proportion of infants who received artificial feeding immediately after birth was 53.6 per cent among Kuwaitis, and 32.3 per cent among non-Kuwaitis.

A review of the number of deaths among the age group 0–2 in the years 1975–76 indicated that gastroenteritis was the main cause of deaths, and the numbers were almost twice as many in Kuwaitis as in non-Kuwaitis.

In the period of March 1978–February 1979, Mostafa *et al.*[8] surveyed feeding and weaning practices in a cross-sectional sample of 966 infants (3–24 months), 580 Kuwaitis, and 386 non-Kuwaitis in high, middle, and low socioeconomic grades. They found that breast feeding was much less practised than bottle feeding by Kuwaitis and non-Kuwaitis, especially among high and middle socioeconomic grades.

At 6 months only about 10.6 per cent of Kuwaitis and 16.3 per cent of non-Kuwaitis of grade I were breast fed compared with 53.1 per cent and 52.5 per cent of grade II, respectively. Humanized milk was much more widely used for artificial feeding (about 85 per cent of Kuwaitis and 64 per cent of non-Kuwaitis) than dried full-cream milk.

The major reasons for not breast feeding were: insufficient milk or milk had dried up (63 per cent grade I; 28.3 per cent grade II); and onset of a subsequent pregnancy (0.9 per cent grade I, 40.8 per cent grade II). A small percentage of mothers gave the contraceptive pill as a reason (3.4 per cent grade I, 2.5 per cent grade II), and some had asked physicians to give them an injection to suppress breast milk.

Oman

A study on 'Diet and Nutrition'[9] was carried out in Nizwa (in the interior of Oman) and Sohar (on the Batinal coastal region) from October 1972 to May 1973, in preparation for the launching of a basic services scheme. Women in 382 households in the two villages were interviewed for eating habits and infant-feeding practices. Several of the families were also observed in detail during home visits.

After delivery, the infant is not put to the breast immediately, because colostrum is believed to be harmful. In the interim (about 3–7 days), prelacteal feeds are offered (honey or molasses alone or in combination with *ghee*, saffron or dates; pepper grass; sugar solution; milk, powdered or goats')—all dripped from the mother's fingers into the baby's mouth, but no feeding bottles are used.

In Sohar, the Koranic teaching is followed to the letter; 90 per cent of male and female infants are breast fed until 2 years of age. Only fifteen mothers stated that they extended breast feeding past this time.

In Nizwa, 16 per cent of mothers reported weaning their babies between 6 and 12 months of age. Weaning at the age of 2 years or later varies according to the sex of the child; 70 per cent of mothers stated that female infants would be weaned at the age of 2 years, and 28 per cent after the second year, compared with 77 per cent and 25 per cent, respectively, for males.

The majority of mothers in Nizwa and Sohar (59 and 55 per cent, respectively) said that if their milk decreased markedly or dried up they would use powdered milk and baby bottle, recently introduced by health personnel or by women who have returned to Oman from Zanzibar. The mode of feeding powdered milk is almost always the bottle rather than the traditional *shali* (eye dropper) or cup. The water used to prepare the powdered milk formula is not boiled. Also, mothers tend to use less than the recommended amount so that the can of milk will last longer.

In 1977, Autret and Miladi[10] visited two MCH centres in Muscat (the capital city) Matrah and Dar Seat. They reported that out of an average of 100 children examined daily in Matrah, ninety were bottle fed before the age of 1 year; two were afflicted with severe marasmus, about twenty had moderate marasmus, and most of the remainder failed to thrive. In Dar Seat, the situation was much more hazardous, because the medical staff distributed free samples of baby foods provided by companies, and advertisement posters were found everywhere in the centre.

Qatar

In 1979, after a visit to the Department of Preventive Medicine at the Ministry of Health, and the children's division of the central hospital, Autret and Miladi[11] noted that the nutritional state of infants and pre-school children bears a striking similarity to the hazardous situation in other Gulf States. The leading causes of nutritional problems in this age group are the declining incidence of breast feeding, and bottle feeding of imported artificial milk and baby foods. Several children afflicted with marasmus associated with gastroenteritis were seen at the children's ward of the central hospital; this was largely because of overdiluted milk formulae, prepared by mothers who lacked knowledge in proper hygiene.

The report deemed it necessary to prohibit commercial advertisement of baby foods, and exert special efforts to promote breast feeding. It also urged that a national survey be carried out to assess the nutrition state in the country, giving special consideration to weaning foods and feeding practices.

United Arab Emirates (UAE)[12]

About 50 per cent of mothers in UAE cease to breast feed their infants before the age of 3 months, and not many of them actually attempt to breast feed. The rapid decline in the incidence of breast feeding appears to be a drastic phenomenon. While 15 per cent of infants under 3 months are exclusively breast fed, 50 per cent are exclusively bottle fed. Wide variations, however, exist among the ethnic groups.

Although bottle feeding was originally introduced to the country by foreigners, now it appears that the native mothers are not interested in breast feeding, primarily because they have a high purchasing power and the desire to spend money on their infants. Since breast milk is a 'free gift' and does not cost money, it is 'not worth much'. Hence, the traditional prestige of breast milk as the 'natural' and 'God-made' best food for the infant has declined tremendously, and the cultural outlook of the new generation of mothers seems to be totally different from that of their own mothers.

Moreover, health workers, particularly physicians, are utilized by commercial companies for advertising artificial baby foods, which abound on the market. Seventeen different varieties were reported in 1978. They are sold to the public at a very high price, often after the expiry date. MCH centres encourage their use as substitutes or supplements for breast milk; at the same time no educational effort is exerted to teach mothers the proper preparation of milk formulae. Baby foods are sold in big boxes to impress the consumers, but the content is usually smaller than the amount stated on the label.

Children's clinics are usually congested with infants under 1 year of age, afflicted with gastroenteritis, dehydration, and varying degrees of PEM, despite the high income of their families and the abundance of protein foods on the market.

In 1978, close to 50 per cent of children who attended child health clinics in Abu Dhabi suffered from gastroenteritis; 30 per cent of infant deaths in the country (April, 1978) were due to gastroenteritis; 90 per cent of them occurred during the first year of life, and 75 per cent in the first 6 months. In view of the magnitude of this problem and its imminent threat to the health and survival of infants, the reporters urged the Ministry of Health to give it the highest priority.

Yemen Arab Republic

The unique social situation in YAR and the rapidly increasing external influences on a previously isolated society undoubtedly played a great role in the infiltration of feeding bottles into the country. Whereas the bottle was unknown before the 1962 revolution,[1] cheap plastic bottles had become available on the market and were increasingly common by 1971, and viewed as a 'symbol of modernity'. The majority of infants in YAR now appear to be bottle fed, and feeding bottles can be purchased in the smallest shops, even in remote villages.[13]

Historically, powdered milk probably first began to enter North Yemen from Aden, the Capital city of South Yemen. Even now many North Yemenis refer to it as *halib al-Adeni* (Aden milk).[13] In 1979, the booming market for artificial baby foods was penetrated by twenty brands of infant formula, about thirty of full-cream milk powder, and seven of evaporated milk; and by appealing pictures of smiling babies on milk tins and brand names such as Nono (Arabic for 'baby').

Although breast feeding is still practised, especially in villages, it is being supplemented sooner with the bottle and stops altogether at an earlier age. In 1972, among mothers attending a CHC in Taiz, 65 per cent had stopped breast feeding by 4 months, and only 16.5 per cent breast fed for a year or more. In 1972–73, in a sample of fifty-four randomly selected women employed by a textile factory in Sana'a, 85 per cent of their infants (1 month to 2 years of age) were bottle fed with or without breast feeding. In the villages this figure dropped to 25 per cent.[1] In 1979, 22 per cent of infants in Al-Qa' had been put on the bottle by 1 month of age, and two-thirds of them by 3 months. Whereas in Udain, 50 per cent of under-five children had been introduced to bottle feeding by 1 month of age, and 75 per cent by 5 months. In the two localities, mothers were found to be practising bottle feeding until rather late ages, often beyond the third year.[13]

Bottle feeding appears to be a major cause of the high prevalence rate of PEM currently observed in YAR, mostly in infants less than 6 months of age. In 1978 the Ministry of Health wrote that almost all infants are breast fed for the first few months, and although the Koran instructs a mother to breast feed for 2 years, women in urban areas are beginning to opt in favour of bottle feeding, with all the associated health hazards.

Summary and conclusions

The Gulf States appear to face now, probably, the most critical bottle-feeding invasion in the world. As modern technology helped to bring about great wealth and unprecedented prosperity to these countries, it introduced also the 'small lethal bomb—the feeding bottle', enhancing the untimely death of young infants and increasing the prevalence of gastroenteritis, cholera, dehydration, and infantile marasmus, often occurring in epidemic waves under the effect of the hot climate.

The uniqueness of breast milk[14] as the 'natural' and 'most appropriate' food for the human infant is gaining increasing emphasis in the light of new information on its nutritive, immunological, emotional, demographic, and economic advantages and a voluminous publication has recently appeared on the subject.[15] The cyclical changes of breast-feeding patterns with the changing ecological context and the image of the breast-feeding mother have been delineated from the preindustrial to the post-industrial phases of socio-economic development.[16,17] Resurgence of breast feeding[18,19] is becoming appreciable in several industrialized countries. The major question for the Gulf States is how to revitalize breast feeding in the immediate future. National policies for the control of baby foods and commercial advertising, psychosocial factors[20] at the root of maternal behaviour, and the changing role of 'mothering', nutrition education, supportive legislation, and social services[21] are urgent matters and priority issues for achieving the task.

Acknowledgements

Special thanks are due to the WHO Office, Eastern Mediterranean Region, particularly Dr R. Cook, for granting access to unpublished reports and data.

References

1. Bornstein, A. The young child in Yemen. *Les Carnets de l'enfance* **28**, 24–42 (1974).
2. Musaiger, A.R.O.A. *A study of food habits in urban and rural areas in Bahrain.* Thesis submitted for partial fulfilment of the MPH degree. High Institute of Public Health, Alexandria University, pp. 10–14 (1977).
3. Autret, M. and Miladi, S. *Report on the food and nutrition situation in Bahrain.* UNICEF Gulf Area Office, Abu Dhabi, UAE (1979). (Unpublished observations.)

4. Gunn, R. A., Kimball, A. M., Pollard, R. A., Feeley, J. C., and Feldman, R. A. Bottle feeding as a risk factor for cholera in infants. *Lancet* **2**, 730–2 (1979).
5. Bin Kadim, H. A. *A study of the characteristics of the first 1000 mothers attending the Family Planning Clinic of the Khormaksar MCH/FP Centre, Aden.* Project: YEA/MCH/001. Directorate of MCH/FP. Ministry of Health, P.D.R.Y. (1977). (Unpublished observations.)
6. Kristianson, B. Bägenholm, G., and Nasher, A. *A pilot study of nutrition and growth of South-Yemeni children* (1978). (Unpublished observations.)
7. MER. *Possible hazards of artificial feeding in Kuwait.* Monthly Epidemiological Report, No. 478, pp. 1–3 (1978).
8. Mostafa, S. A. and Nuwayhed, H. Y. *Milk feeding and weaning practices among infants in Kuwait.* Unit of Nutrition, Preventive Health Section, Department of Public Health and Planning, Ministry of Public Health, Kuwait (1979). (Unpublished observations.)
9. UNICEF Gulf Area Office. *Diet and Nutrition in Nizwa and Sohar.* Sultanate of Oman. pp. 79–103 (1973).
10. Autret, M. and Miladi, S. *Report on the food and nutrition situation in the Sultanate of Oman.* UNICEF Gulf Area Office, Abu Dhabi, UAE (1979). (Unpublished observations.)
11. Autret, M. and Miladi, S. *Report on the food and nutrition situation in Qatar.* UNICEF Gulf Area Office, Abu Dhabi, UAE. (1979). (Unpublished observations.)
12. Autret, M. and Miladi, S *Report on the food and nutrition situation in the United Arab Emirates.* UNICEF Gulf Area Office, Abu Dhabi, UAE (1978/1979). (Unpublished observations.)
13. Greiner, T. *Background paper on breast and bottle feeding.* (Unpublished observations.)
14. Jelliffe, D. B. and Jelliffe, E. F. P. (eds.) Symposium: the uniqueness of human milk. Reprint. *Am. J. Clin. Nutr.* **24**, 968–1024 (1971).
15. Jelliffe, D. B. and Jelliffe, E. F. P. (eds.) *Human milk in the modern world.* Oxford University Press, p. 500 (1978).
16. Jelliffe, D. B. Epidemiology of under nutrition. In *Nutrition in the community* (ed. D. S. McLaren). J. Wiley, London, pp. 87–99 (1976).
17. Pellet, P. L. Commentary: marasmus in a newly rich urbanized society. *Ecol. Food Nutr.* **6**, 53–6 (1977).
18. Péchevis, M. Elements for a policy promoting breast feeding. *Assignment Child.* **32**, 33–49 (1975).
19. Sjölin, S. Present trends in breast feeding. *Curr. Med. Res. Opinion* **4**, Suppl. 1, 17–22 (1976).
20. Harfouche, J. K. Psychosocial aspects of breast feeding, including bonding. *Food Nutr. Bull.* **2**, 2–6 (1980).
21. Savané, M. A. Yes to breast feeding, but . . . how? *Assignment Child.* **49/50**, 81–7 (1980).

Index